PEERING
CARRIER ETHERNET NETWORKS

PEERING
CARRIER ETHERNET NETWORKS

SACHIDANANDA KANGOVI

AMSTERDAM • BOSTON • HEIDELBERG • LONDON
NEW YORK • OXFORD • PARIS • SAN DIEGO
SAN FRANCISCO • SINGAPORE • SYDNEY • TOKYO
Morgan Kaufmann is an imprint of Elsevier

Morgan Kaufmann is an imprint of Elsevier
50 Hampshire Street, 5th Floor, Cambridge, MA 02139, United States

Library of Congress Cataloging-in-Publication Data
A catalog record for this book is available from the Library of Congress

British Library Cataloguing-in-Publication Data
A catalogue record for this book is available from the British Library

ISBN: 978-0-12-805319-5

For information on all Morgan Kaufmann publications
visit our website at https://www.elsevier.com/

Working together
to grow libraries in
developing countries

www.elsevier.com • www.bookaid.org

Publisher: Todd Green
Acquisition Editor: Brian Romer
Editorial Project Manager: Ana Claudia A. Garcia
Production Project Manager: Priya Kumaraguruparan
Cover Designer: Matthew Limbert

Typeset by TNQ Books and Journals

To

Sita, Shreya, and Rob;

My parents—Bapaji and Shanthu;

My brother Sinu and his family;

Sita's sisters and brother and their families; and

My friend Gajanana.

CONTENTS

Contents ix

LIST OF FIGURES

LIST OF TABLES

ABOUT THE AUTHOR

Sachidananda Kangovi is an Enterprise Architect, currently consulting at AT&T. He has over 20 years of experience in IT systems related to operations and business support systems in Telecom. While working as a Distinguished Engineer at Comcast, he developed a state machine called Service Linked Multistate System for which he was awarded many US and international patents. He was also VP for Telecom Systems at SCSL, which was a global IT company and Director of Systems Engineering at ADC/CommTech.

Kangovi is also a former Adjunct Professor in Keller School of Management at DeVry University where he taught courses on Business Intelligence and Data Analysis, Advanced Program Management, and Managing Software Development Projects.

Kangovi has a Ph.D. in Engineering from Rutgers University and an Executive MBA from Universitas 21—a consortium of 21 leading global universities. He did his MS from Indian Institute of Science and his BS from Jabalpur University.

PREFACE

About this Book

It took over 100 years, starting in 1876, for telephone networks to develop, grow, and mature. Compared to that, data networks, starting in 1969, took 50 years. Wireless networks, on the other hand, beginning around 1990, took just about 25 years to provide the present state in mobility. During large part of its existence telephone network, known as public-switched telephone network (PSTN), was a regulated monopoly to become the largest network that mankind had built, whereas development of data and wireless networks took place in an atmosphere of extreme competition. Today, data and wireless networks have grown to such an extent that the telephone network is becoming obsolete.

As a consequence of all these rapid changes and competition, in the present environment, various technologies exist side by side giving us diversity of access. Diversity of access is a wonderful thing for expanding the customer base. However, the existence of multiple technologies has also created a diversity in backhaul, which is not a desirable thing, because it adds to complexity and to the total cost of ownership (TCO) for service providers and operators.

The good news is that the data network has emerged as the most dominant network with Transmission Control Protocol/Internet protocol (IP)/Ethernet as the de facto stack of protocols and the advances in the Ethernet technologies have led to its applicability from local to global networks. This is also enabling higher bandwidth, low frame delay (latency), low frame delay variation (jitter), higher reliability, and scalability of the networks. These advances are leading to convergence in backhaul which is good for reducing complexity and TCO. Furthermore, the growth of voice over IP/voice over long-term evolution is converting voice into another data application and thus eliminating the need for PSTN.

Besides reducing complexity and TCO, these developments also have tremendous potential to support, at a reasonable cost, new and emerging applications such as virtual reality (VR), Internet of things (IoT), CPS, 3-D video, and cloud applications.

In order to leverage Ethernet technology and to develop specifications to standardize Ethernet services and make them carrier grade, a forum called Metro Ethernet Forum (MEF) was formed in 2001 consisting of representatives from information and communications technology industry, universities, and R&D

organizations. These standardized and carrier-grade services are known as Carrier Ethernet (CE) services, and the associated data network came to be known as CE Network (CEN). These CENs are now growing rapidly, which has led to the need to peer CENs in order to provide services to off-net customers. And, that is the subject of this book.

Organization of the Book

This book attempts to trace the evolution instead of exclusively emphasizing on the current state of the technology. Hopefully, this will make the book more interesting to read besides underscoring the simple origins of almost all the complexities in this technology. Each chapter also begins with a relevant quotation as an "icebreaker."

Chapter 1 surveys the landscape vis-à-vis PSTN, data networks, hybrid–fiber–coaxial networks of cable operators and wireless networks. This survey covers the origins of many important building blocks and identifies significant trends such as diverse access methods in the local loop and need for convergence in the backhaul, and the need for higher bandwidth and performance. The chapter also interleaves the technological developments with the changing legal and business environment.

Chapter 2 is all about shaping of data networks. It describes the Open Systems Interconnection (OSI) seven-layer model and the Internet service provider–centric architecture blueprint from National Science Foundation that later morphed in to a geography-centric architecture. This chapter also describes the convergence in backhaul by the emergence of Ethernet as the most popular protocol first at layer 2 and subsequently at layer 1 as well of the OSI seven-layer model. Lastly, the chapter covers IP Multimedia Subsystem platform which is transforming voice into another data application.

Chapter 3 expands on the conclusions of first two chapters and describes the Ethernet evolution to achieve higher bandwidths and performance over longer distances. The implementation details of the Ethernet technology in hardware at chip and device levels are also covered in this chapter.

Chapter 4 introduces CENs and describes the formation of the MEF for standardizing Ethernet services, and specifying quality of service (QoS) and service management. This chapter describes different Ethernet service types, service attributes, and parameters and presents the definitions of CE, CENs, and associated terminology including User Network Interface and Ethernet virtual connection. The chapter also describes service operation, administration and management (SOAM) functions for fault and performance monitoring in order to ensure that QoS is in compliance with the service-level agreements.

The rapid growth of CENs has made it necessary to peer CENs belonging to different operators in order to provide services to off-net customers. This is the subject of Chapter 5. The chapter covers E-Access service type including architecture, attributes, parameters, and terminology associated with it. The chapter presents definitions of external network–network interface and operator virtual connections associated with E-Access service type. The role of bridging techniques and tags, particularly S-tags, are further explained in this chapter because of the critical role they play in peering CENs. The chapter describes how QoS is delivered on peering CENs by coordination of class of service (CoS), performance parameters, and policing of bandwidth profile by two-rate three-color model based on "token-bucket" algorithm. Description in this chapter also includes SOAM over peering CENs.

Chapter 6 describes business-to-business (B2B) transactions that the peering of CENs requires between service providing operator and access providing operator. In communication industry, this B2B transaction is called access service request (ASR), and this chapter provides a brief descriptions of various ASR forms and fields therein and the values they take.

Chapter 7 describes the architectural framework and functions of operations and business support systems (OSS/BSS). These are a large and complex group of IT applications which provide automation or mechanization of multitude of activities prior to and following an ASR. The chapter also describes the Next Generation Operations Systems and Software specification that includes Enhanced Telecom Operations Map Framework for coordination of marketing, infrastructure, customers, products and services, resources, operations, service assurance management, physical and virtual inventory, and billing. The chapter then covers additional functionalities, which a well-designed, robust, secure, and redundant OSS/BSS system must support. Many features of OSS/BSS systems including support for self-care, help desk outsourcing, network function virtualization, and software-defined networking are also described in this chapter.

Chapter 8 covers the taxonomy of customers and their applications. It describes how the taxonomy is useful in understanding applications that customers need, use, and pay for. The chapter then presents the mapping of the application-specific performance objectives to MEF-defined standard CoS performance objectives and performance tiers. This mapping is crucial to standardizing CE services in CENs and peering CENs. The chapter then covers the applicable CE services including E-Access service for peering CENs to meet the network functionality needed by customer applications. Examples of network functionality include IP backhaul, mobile backhaul, streaming and switched video transport, site-to-site connectivity, connection for cloud computing services, and network connectivity for emerging applications such as IoT, cyber-physical systems, and VR. The chapter then dwells, briefly, on a process to convert information about customer applications and topology into a design for a CE service based on CENs and peering CENs.

Chapter 9, the last chapter of the book, examines some of the next steps needed in the Ethernet technology and peering CENs and also in OSS/BSS systems to meet the growing demands for high-bandwidth and high-performance CENs and peering CENs to support current and emerging applications.

Miscellaneous Items

It is important to point out couple of items encountered during the writing of this book. First item relates to plethora of acronyms in the information and communications field. Some of these are included in a glossary. One will notice that, due to rapid growth of this field, there are some acronyms that are overloaded i.e. have multiple meanings. For example CE could mean carrier Ethernet services as well as customer edge devices. Similarly CM could mean cable modem or it could also mean color mode. These acronyms are based on common usage and some have even entered standards/specifications. In this book no attempt has been made to correct the multiple meanings as that would require larger industry effort.

The second item relates to the word "data". Although various English dictionaries describe "data" as plural but there is considerably controversy about if "data" is a singular, uncountable noun, or should be treated as the plural of the now-rarely-used datum. The description in https://en.wikipedia.org/wiki/Data_(word) states that the debate over appropriate usage of "data" word continues, but "data" as a singular form is far more commonly used. It is a strange situation - the world is awash in data and yet English language has not yet unequivocally decided if "data'" is singular or plural! Due to the fact that we cannot settle this debate ourselves, we have followed the common practice and not strict definition per dictionary and treated "data" as singular.

Third item relates to specifications and standards. Some of them are evolving and will go through changes and some new standards will emerge with time. In view of this, the goal in this book has been to provide a view of the building blocks and basic concepts leading up to the development of peering CENs and not on reproducing all the details in these specifications and standards. It is hoped that this approach will, in turn, help in understanding the specification/standards better. The book also includes other areas like operations and business support systems and access service request which are also important to peering of CENs, thus providing a holistic view of this field.

Audience of the Book

The audience for this book includes those with intermediate to advanced knowledge of networking and telecommunications with interest in the Ethernet technology, CENs, peering CENs, and OSS/BSS. These readers could be from telecommunications companies, multiple system operators, system integrators, network equipment vendors, OSS/BSS system vendors, universities, industry forums, and standards organizations.

ACKNOWLEDGMENTS

It is impossible to write a book without help from many. I wish to begin with thanking Brian Romer, Senior Acquisition Editor, because it all started with him. It is Brian who organized a panel of experts to review the initial outline proposal of this book. I express my gratitude to him for his support and cooperation.

I wish to thank the proposal reviewers: Larry Samberg, member of Technical Committee of Metro Ethernet Forum (MEF) and Twinspruces Consulting; Lars Dittman, Professor, University of Denmark, Lyngby, Denmark; Luis Almeida, Associate Professor, University of Porto, Portugal; and Paulo Monteiro, Associate Professor, University of Aveiro and Researcher at the Institute de Telecomunicacoes, Portugal. Their feedback and valuable suggestions encouraged me to go ahead with this book.

I am especially grateful to Larry Samberg who reviewed the manuscript of the book. Larry took time from his busy schedule to offer in-depth comments and advice. He was always available to offer clarifications in response to my queries. He researched material and even contacted authors of MEF specifications. He was generous to share with me the draft of an MEF specification that he himself has authored. I have vastly benefited from his nearly four decades of experience in this field and his close association with the MEF. It is worth noting that Larry founded an independent Ethernet bridge company as early as 1987. Larry's review has, without doubt, enhanced the value of this book.

I am thankful to Anna Tavora Enerio, Director of Marketing, PARC, a Xerox Company, for permission to reproduce the diagram shown in Fig. 3.1A. My thanks are also due to Kenneth Dilbeck, VP, Collaboration R&D at TM Forum, for permission to reproduce Enhanced Telecom Operations Map and SID frameworks shown in Figs. 7.6 and 7.7, respectively. I thank Riick, the author of the diagram shown in Fig. 2.9, for making it available under GNU Free Documentation license on Wikipedia. Carolyn Potts of Digibarn; Hop Wechsler, Manager at Elsevier Permissions Helpdesk; and Susan Mulhern of PARC researched the origin of the diagram shown in Fig. 3.1B. Unfortunately, it could not be ascertained. Nevertheless, I wish to appreciate efforts of these individuals.

I wish to thank Morgan Kaufmann, Imprint of Elsevier, Inc., for agreeing to publish this book.

I wish to express my gratitude to Amy Invernizzi and Ana Claudia Garcia, Editorial Project Managers, for their cooperation and efficient coordination throughout the process of writing this book. My thanks are also to Mathew Limbert, Cover Designer, for the excellent cover design and Priya Kumaraguruparan, Production Project Manager, for expertly managing the production of this book.

Finally, I wish to express my sincere gratitude to countless intelligent, insightful, and supportive individuals with whom I have worked in some of the great companies. They helped me understand the fast developing technologies in the communications and information field.

Sachidananda Kangovi
July, 2016

GLOSSARY

AAA Authentication, authorization and accounting
APO Application performance objective
APS Automatic protection switching
ASIC Application-specific integrated circuit
ASR Access Service Request
ATIS Alliance for Telecommunications Industry Solutions, Inc.
B-DA Backbone-destination address
BDW Bandwidth
BID Bridge ID
BPDU Bridge protocol data unit
B-SA Backbone-source address
BSS Business support system
B-VID Backbone-virtual LAN identifier
CAC Connection admission control
CBS Committed burst size
CCM Continuity check message
CDMA Code-division multiple access
CE Carrier Ethernet (it is also used for customer edge)
CEN Carrier Ethernet network
CE-VLAN Customer edge-virtual local area network
CF Coupling Flag
CGMII 100-gigabit MII
CIR Committed information rate
CLCI Common language circuit identifier
CLEI Common language equipment identifier
CLFI Common language facility identifier
CLLI Common language location identifier
CM Cable modem
CM Color Mode
CMTS Cable modem terminating system
CO Central office
CoS IA Class of service implementation agreement
CoS Class of service
CPE Customer premises equipment
CPO CoS performance objective
CPS Cyber-physical system
CSCF Call signal control function
CSMA/CD Carrier sense multiple access/collision detection
DCE Data circuit terminating equipment
DEI Drop eligibility identifier
DIX Consortium of Digital Equipment Corporation, Intel and Xerox Corporation
DLF Destination lookup failure
DMT Discrete multitone
DOCSIS Data over cable system interface specification
DSAP Destination service access point
DSCP DiffServ code point
DWDM Dense wavelength-division multiplexing

EBS Excess burst size
EIP Ethernet interconnect points
EIR Excess information rate
E-LMI Ethernet local management interface
eMTA Embedded multimedia terminal adapter
EMUX Ethernet multiplexer
ENNI External Network-to-Network Interface
eNodeB Evolved NodeB
EPC Evolved packet core
EP-LAN Ethernet private LAN
EP-Line Ethernet private line
EP-Tree Ethernet private tree
eTOM Enhanced telecom operations map
EUSA End User Special Access
eUTRAN Evolved universal terrestrial radio access network
EVC Ethernet virtual connection
EVP-LAN Ethernet virtual private LAN
EVP-Line Ethernet virtual private line
EVP-Tree Ethernet virtual private tree
FAB Fulfillment, assurance and billing
FCAPS Fault, configuration, accounting, performance and security
FCS Frame check sequence
FD Frame delay
FDB Forwarding database
FDMA Frequency-division multiple access
FDR Frame delay range
FDX Full duplex
FLR Frame loss ratio
FPGA Field-Programmable Gate Array
FTTx Fiber To The x (here x is replaced by C for cabinet, H for home, B for building etc.)
GARP Generic Attribute Registration Protocol
GBIC Gigabit Ethernet interface
GIS Geographical information system
GMII Gigabit MII
GNU Generally Not UNIX
GPON Gigabit passive optical network
GPRS Global packet radio system
GSM Global system for mobile communication
HDX Half duplex
HFC Hybrid fiber-coaxial
HLR Home location register
HSS Home subscriber server
ICT Information and communications technology
IEEE Institute of Electrical and Electronics Engineers
IFDV Interframe delay variation
IMS IP multimedia subsystem
IMTS Improved mobile telephone service
IoT Internet of things
IP Internet protocol
IPTV Internet Protocol television
I-SID Service ID
ISP Internet service provider
ITU International Telecommunication Union

ITU-T ITU—Telecommunication Standardization Sector
J2EE Java 2 Platform Enterprise Edition
L2CP Layer 2 control protocol
LACP Link Aggregation Control Protocol
LAG Link Aggregation Group
LAN Local area network
LBM Loopback message
LBR Loopback response
LC Lucent Connector
LLC Logical link control
LOS Level of service
LTE Long-term evolution
LTM Link trace message
LTR Link trace response
MAC Media access control
MACD Move, add, change and disconnect
MAE Metropolitan area exchange
MAN Metropolitan area network
MAU Medium attachment unit
MDF Main distribution frame
MDI Medium-dependent interface
ME Maintenance entity
MEF Metro Ethernet Forum
MEG Maintenance entity group
MEP Maintenance end point
MFD Mean frame delay
MGCF Media gateway control function
MGW Media gateway
MII Medium-independent interface
MIP Maintenance intermediate point
MME Mobility Management Entity
MMF Multimode fiber
MP2MP2 Multipoint-to-multipoint
MRCF Media resource control function
MSAG Master street address guide
MSO Multiple system operators
MTSO Mobile telephone service operator
MTTR Mean time to restore
NAP Network access point
NC Network channel
NCI Network channel interface
NFV Network function virtualization
NGOSS Next generation operations systems and software
NIC Network interface card
NID Network interface device
NTE Network termination equipment
O EP OVC end point
OBF Ordering and Billing Forum
OFDMA Orthogonal frequency-division multiple access
OLT Optical line termination
ONT Optical network terminal
OSI Open systems interconnections
OSS/BSS Operations and Business Support Systems
OTT Over-the-top

OVC Operator virtual connection
P2P Point-to-point
PB Provider bridge
PBB Provider backbone bridge
PBB-TE Provider backbone bridge—traffic engineering
PCM Pulse-code modulation
PCP Priority code point
PCRF Policy charges and rules function
Pd Frame delay performance percentile
PGW Packet gateway
PHY Physical layer
PLS Physical layer signaling
PMA Physical medium attachment
PNF Physical network function
PON Passive optical network
Pr Specific range of the frame delay performance used in FDR
PRC Primary reference clock
PSTN Public switched telephone network
PT Performance tier
Pv Frame delay variation performance percentile
QAM Quadrature amplitude modulation
QoS Quality of service
RAN Regional area network
RG Residential Gateway
RMII Reduced MII
ROADM Reconfigurable optical add-drop multiplexer
RRH Remote radio head
RSTP Rapid Spanning Tree Protocol
Rx Receive
SALI Service address location information
SAP Service access point
SAT Source address table
SC Standard Connector
SDN Software-defined network
SECNCI Secondary network channel interface
SFD Start frame delimiter
SFP Small-form factor pluggable
SGW Serving gateway
SID Shared information/data
SLA Service-level agreement
SLIMS Service-linked multistate system
SLO Service-level objective
SLS Service-level specification
SMF Single-mode fiber
SMS Short message service
SNAP Subnet access protocol
SOA Service-oriented architecture
SOAM Service Operation, Administration and Management
SOAP Simple object access protocol
SSPA Source service access point
STP Spanning Tree Protocol
SWC Serving wire center
T A time interval that serves as a parameter for service-level specification
TAM Telecom applications map

TCI Tag control information
TCO Total cost of ownership
TCP Transmission control protocol
TDM Time-division multiplexing
TDMA Time-division multiple access
TMN Telecommunication management network
TNA Technology neutral architecture
TOM Telecom operations map
TOS Type of service
TPID Tag protocol identifier
Tx Transmit
UDDI Universal description, discovery, and integration
UNI User network interface
VLAN Virtual local area network
VLR Visitor location register
VM Virtual machine
VN Virtual network
VNF Virtual network function
VoIP Voice over IP
VoLTE Voice over LTE
VR Virtual reality
WAN Wide area network
WRC World radio conference
WSDL Web service definition language
xDSL x Digital subscriber line (here x could be A for ADSL and V for VDSL etc.)
XGMII 10-gigabit MII
XLGMII 40-gigabit MII
Δt A time interval much smaller than T

INTRODUCTION

What use could this company make of an electrical toy?

Remarks by President of Western Union, upon turning down Gardiner Hubbard's offer to sell him the patent rights to telephone, in 1876.

Near Princeton in New Jersey, there is Delaware & Raritan (D&R) canal, busy North-East Corridor train tracks, and US Route 1 in close proximity. This place is also close to some of the busiest airports in the nation. These different modes of transportation represent our quest to gravitate toward higher speeds, more flexibility, and increased ease of use while improving reliability and reducing costs. Any new method that improves on these aspects obsoletes earlier methods as is the case with the D&R canal.

What is true in public and goods transport systems is also true with communication transport systems. There is constant push for higher throughput, low frame delay (latency), low frame delay variation (jitter), higher reliability, and lower total cost of ownership (TCO). This push is leading to new developments, fairly rapidly, especially in last few years. In order to understand these new developments, it is important to look at what came before because many important building blocks such as networks, switching, and multiplexing evolved over time.

This chapter provides an overview of telephone networks, data networks, video [hybrid fiber-coaxial (HFC)] networks, and wireless networks identifying the origins of many important building blocks in the evolution.

1.1 Telephone Networks

Telephone networks trace their origin to the Bell Patent Company formed by a business agreement between Alexander Graham Bell, Gardiner Hubbard, and Thomas Saunders in 1875. This funded Bell's research to invent a device to transmit multiple telegraph signals over a single line. However, the research instead led to the invention of the telephone[1] in 1876 as shown in Fig. 1.1. This was a *ring-down circuit* where one person used the transmitter, and another person used the receiver. In other words, it was a one-way voice communication, and there was no dialing or ringing.

Peering Carrier Ethernet Networks. http://dx.doi.org/10.1016/B978-0-12-805319-5.00001-0

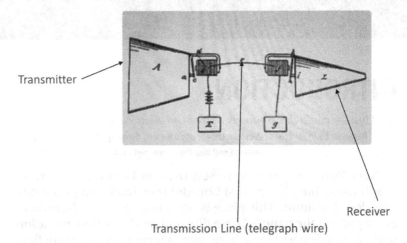

Transmitter

Receiver

Transmission Line (telegraph wire)

Figure 1.1 Diagram from Alexander Graham Bell's patent application.

After Western Union turned down the patent rights to use the telephone, the Bell Telephone Company was formed in 1877 to manufacture telephones under the Bell patent which was granted to Alexander Graham Bell effective March 7, 1876. This turned out to be one of the richest patents in the history.

The one-way phone soon evolved in to two-way voice communication which allowed users on both ends to transmit voice. Bell Telephone Company would lease two telephones one for each end. This resulted in a *two-way party line* as shown in Fig. 1.2.

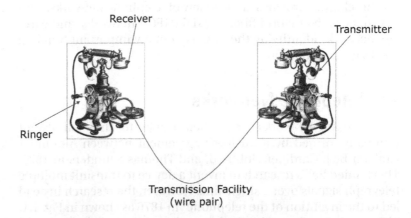

Receiver

Transmitter

Ringer

Transmission Facility
(wire pair)

Figure 1.2 Point-to-point connection (two-way party line).

This arrangement limited the usefulness of the telephone because as number of customers grew it required large number of wire pairs to interconnect telephones. For example, if N number of customers wanted to connect with each other on party line, then it would require $N \times (N-1)/2$ pairs of wires to interconnect all the telephones. Just to illustrate, if N is equal to 20, then we would need 190 wire pairs to interconnect all 20 telephones of these 20 customers.

This issue was solved in 1878 by establishing exchanges with manual operators. These exchanges were run by independent companies under license from Bell Telephone Company. These independent companies later became Bell Operating Companies (BOCs). Fig. 1.3 shows the schematic of a typical manual exchange.

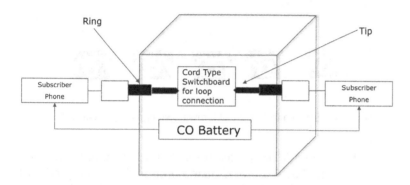

Figure 1.3 Schematic diagram of manual switching.

An out-of-court settlement with Western Union in 1879 recognized Bell's patent rights and required Western Union to stay out of telephone business. It also transferred all of Western Union's telephone facilities and equipment to Bell Telephone Company including 56,000 telephones in 55 cities. This greatly increased Bell's footprint. Soon Bell acquired Western Electric to manufacture telephones for all Bell licensees. This also opened up the need for intercity communication or long distance telephone services.

In 1884, first long distance telephone service was offered between Boston and New York City and in 1885, American Telephone and Telegraph Company (AT&T) was established as a subsidiary to provide interexchange services (this is the old AT&T and not the new one). All local exchanges were required to connect to AT&T for long distance services. This soon made AT&T richer than the parent Bell Telephone Company, and so all assets were transferred to AT&T.

The next big development in telephone communication took place in 1888 when the first electromechanical switch was developed by Almon Strowger, an undertaker. He patented[2] it in 1891. He was motivated to develop this switch because he suspected that the local manual telephone exchange operator, who was the wife of a competitor undertaker, was sending calls meant for him to her husband, and as a result, he was losing business. Fig. 1.4

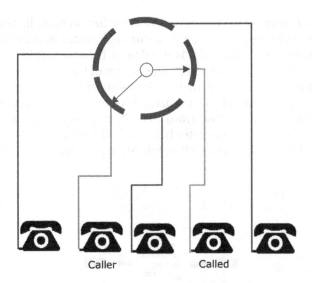

Figure 1.4 Schematic diagram of Strowger electromechanical switch.

shows the schematic of the electromechanical switch. This design has been refined over the years leading up to the development of softswitches, but the principle remains the same which is to connect telephone calls from one phone line to another.

The simple pair of copper wire running from the customer's premise was called local loop. This local loop physically connected the telephone to a switch housed in a central office (CO). The connection between COs is called trunk. As early COs started to interconnect, large amount of copper wire was installed to support the traffic. Soon cities began to face problems from these wires strung from poles. In some places, people could not see the sky. This situation was further accentuated by the expiration of Bell's patent in 1894. Independent companies rushed to compete. By 1902, more than 4000 independent companies had entered the field. This led to many inefficiencies as customers had to lease multiple telephones to connect with users on other company's telephone networks. The old AT&T started to counter the competition by advocating telephone service as a natural monopoly and also buying up the competition.

This problem of excessive copper wires connecting COs was solved by a technology innovation called multiplexing. The first analog multiplexing in telephony was developed by George Owen Squirer[3,4] in 1910. This multiplexing technique has been refined over the years, but the principle remains the same of combining many low speed channels into one high-speed channel. Ironically, in a way, multiplexing was the goal of the original research of

Alexander Graham Bell to send four signals on a single telegraph line that led to the invention of telephone.

In 1912, the Justice Department brought an antitrust suit against the old AT&T. This was settled in 1913 by which the old AT&T agreed to limit its size to 80% of all customers and 40% of the geographical area of the United States. This was the first antitrust suit against the old AT&T, and the settlement was called Kingsbury Commitment. The government, however, agreed with AT&T that the telephone service operated more efficiently as a monopoly, and this was reflected in two legislations. The first one called the Mann–Elkins Act of 1910 placed telephone industry under Interstate Commerce Commission, and the second legislation called the Graham Willis Act of 1921 recognized Bell system as a national resource allowed it to operate as a natural monopoly. These two acts converted pure competition in to a regulated monopoly leading up to Federal Communication Act of 1934 which established Universal Service as the most important goal of the telephone industry. Bell System was allowed to function as a monopoly with price regulations.

In the meanwhile, there was another technical issue that needed attention. This issue was related to the fact that our speech is analog in nature in the range of 0–4000 Hz. The transmitter converted the voice to an electrical signal and put this signal on the copper wire. On the receiver end, this electrical signal was reconverted to the acoustic signal by a diaphragm. The problem was the noise introduced by static on the line. Any amplifiers between the two ends also amplified noise resulting in garbling. Various techniques were developed that successfully dealt with noise. However, the technique of converting the analog voice signal to digital signals made the best sense due to several reasons. In addition to noise reduction, digital signals opened door for other benefits such as increased bandwidth, transmission over longer distances, and processing by digital circuit components which are cheaper and can be compactly produced on a chip. This technique was based on first passing analog voice signal through a filter to remove any signals above 4000 Hz and then sampling the signal 8000 times per second which is twice the voice frequency. This sampling rate was defined by Nyquist theorem developed[5] in 1928. This sampling rate eliminates any aliasing. Amplitude of each of these 8000 samples is represented by 8 bits. This results in 8000 × 8 or 64,000 bits per second or 64 kbps of data. This technique is called pulse-code modulation and is the most common method of digitizing an analog voice signal. Soon after Nyquist published his results, Hartley leveraged it to construct a relation between achievable line rate and number of pulses that could be put through a medium. Hartley's equation, however, did not show exactly how noise influenced this line rate.

In 1948 there was another major breakthrough when Shannon's Information theory[6] enabled inclusion of effects of noise and led to the Shannon–Hartley equation. This provides a relation between bandwidth capacity, maximum frequency, and signal-to-noise ratio of a medium as shown below:

$$C = W.\log_2 (1+S/N) \tag{1.1}$$

here C is the bandwidth in bits per second (bps), W is the maximum frequency that the medium can support, and S/N is the signal-to-noise ratio. Since a copper wire commonly used in telephone line could support frequencies much higher than 4000 Hz required for voice, a system was first developed to multiplex 24 such voice channels on to a copper wire using a frequency of 96,000 (24×4000) Hz. The resulting bandwidth was 1.536 Mbps (64 kbps $\times$ 24). Each sample also needed 1 bit for framing to mark the beginning of each frame. This added a total of 8 kbps to the bandwidth for 8000 samples, resulting in a line rate of 1.544 Mbps. The concept of multiplexing based on this theory is shown in Fig. 1.5.

Early Trunk Lines

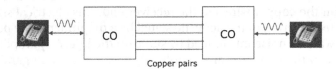

Copper pairs

Digital Multiplexed T1 Trunk

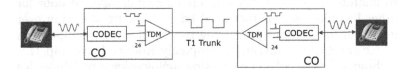

Figure 1.5 Multiplexing in telephone network.

As the popularity of telephones increased, it became necessary to organize the telephone infrastructure into

1. components of local loop including customer premises equipment (CPE), inside wire, line protector, drop wire, terminal block, distribution, service junction, lateral feeder, and main feeder;
2. components of a CO or a serving wire center including cable vault, main distribution frame, transmission equipment, signaling

equipment power supply, equipment for switching, and interoffice transmission; and

3. components of a tandem office including equipment for transmission to and from end offices and routing to end offices. Fig. 1.6 shows organization of the telephone infrastructure.

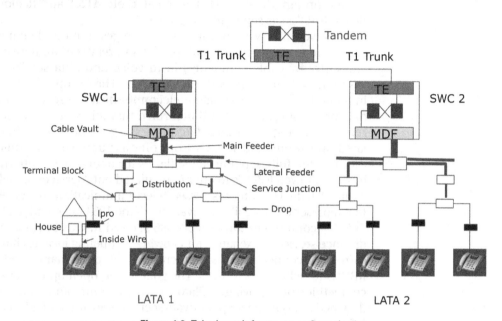

Figure 1.6 Telephone Infrastructure Organization.

The wire pair in the local loop is insulated by plastic and twisted together to reduce cross talk. These twisted pairs are bundled in protective cable sheath. Wire range from 26 gauge (0.016 inches in diameter) to 19 gauge (0.036 inches in diameter). Bundled cables range from six pairs for customer drop to 3600 pairs in feeder. Currently, fiber optics is being introduced in local loop, but copper pairs remain the most common transmission media.

With these building blocks in place, telephone networks expanded not just in the US but around the globe to provide telephone service to millions of customers. Protected as a virtual monopoly to provide universal service, old AT&T was controlling every aspect of this service. But changes were on the horizon.

Change came in the form of microwave transmission. This allowed for connecting COs without wires and made old AT&T's long distance business vulnerable to competition. The second change came from the court and Federal Communications Commission (FCC) rulings that opened up the hold that old AT&T had on the type of equipment that customers could connect to its telephone network. These interconnection rulings and common carrier decisions allowed competition in the CPE and long distance services.

But more changes were to come. The antitrust suit of 1974 led to the famous Modified Final Judgment (MFJ) by Judge Harold Greene on August 24, 1982. This MFJ led to the divestiture of Bell System on January 1, 1984. This created old AT&T and Regional BOCs (RBOCs) as separate companies.

However, this was not the end of changes. Profound changes came due to emergence of computers and related data networks and entry by cable companies in to voice and data services in addition to their traditional video services. The Computer Inquiry initiated by FCC differentiated telecommunication services from data processing services and first ruled that telecommunications services are regulated and data processing services are deregulated and prohibited RBOCs from offering data services unless they created fully separate subsidiaries for data services but then in 1985 allowed RBOCs to offer deregulated data services as well.

In 1996 the US Congress passed Telecommunications Act of 1996 that represented a significant overhaul of the 1934 Act creating FCC. This Telecommunications Act of 1996 allowed RBOCs to enter long distance service in exchange for allowing competing local exchange carriers to connect their equipment to local loops maintained by RBOCs to deliver the services. The goal was to open up markets to competition by removing regulatory barriers to entry, and the act was designed to accelerate rapid private sector deployment of advanced information technologies and services. It did hasten the spread of broadband services, but it also created many service providers, more than what market forces could sustain. This led to the collapse of the dot-com bubble in 2000. The emergence of mobile services was another game changer. Together, these two developments led to consolidation. This time around and in an ironic twist, an RBOC called Southwestern Bell Corporation (SBC) bought old AT&T on November 18, 2005, and changed its name from SBC to AT&T Inc. This is the new AT&T. It has three other RBOCs join the company. Similar consolidations took place among other RBOCs as well.

Now in 2014, because voice is increasingly becoming another data application and the lines of separation between voice and data are vanishing, the FCC chairman has allowed trials on scaling down time-division multiplexing (TDM)–based networks that form the backbone of the public switched telephone network (PSTN). In other words, within about 140 years, PSTNs grew to become the largest network that mankind had built to being torn down and replaced by data networks. However, it gave us many building blocks that are profound and without these data networks or mobile networks would not have been possible. This brings us to our next section on data networks.

1.2 Data Networks

There is no reason anyone would want a computer in their home.
Remarks by Ken Olson—President Digital Equipment Corporation in 1977.

Large standalone mainframe computers evolved in the 1960s popularized by IBM. They were followed by minicomputers primarily from Digital Equipment Corporation (DEC). Soon there was a desire to network them. As a result, RAND Corporation was awarded a project by Advanced Research Project Agency (ARPA) to develop a strategy to network these computers. In 1964 RAND proposed a strategy that was based on dividing data message from the computer into datagrams or packets and each packet having the address of its destination. Each packet would travel independent of other packets, and at the destination, all packets would be assembled in proper order to retrieve the original data message. This strategy was based on the work done by Leonard Kleinrock[7], Paul Baran,[8] and Donald Davies,[9] and it was Davies who called it "packet switching." This term was quickly adopted. It was radically different from the circuit-switching approach used in telephony where a dedicated connection is established between the source and destination for the duration of the call and then at the end of the call that connection is torn down.

Based on this packet-switching strategy, ARPA issued a request for proposal (RFP) in 1968 for creation of such a packet-switching network and labeled the proposed network as ARPANET. In January 1969, a contract was awarded to a company called BBN, and it delivered a switch called Interface Message Processor (IMP) by December of 1969. It was built on a Honeywell DDP 516. The functions of an IMP included dial-up, error checking, retransmission, routing, and verification of messages on behalf of the host computer. These IMPs were installed at UCLA (University of California at Los Angeles), Stanford Research Institute, University of California in Santa Barbara (UCSB), and University of Utah in Salt Lake City. Four different varieties of computers were connected to these IMPs. In UCLA it was a Sigma 7; in UCSB it was IBM 360-75; in SRI it was an SDS-940; and at University of Utah it was a DEC PDP-10. These IMPs were in turn connected by 50 kbps leased lines according to the BBN report number 4799, page III-16, April 1, 1981, to complete the ARPANET as shown in Fig. 1.7. A leased line is a phone line dedicated to data transmission. The sending and receiving devices were digital, but the line in the middle was leased phone line. These lines belong to telephone companies and sold to customers as dedicated resource available continuously.

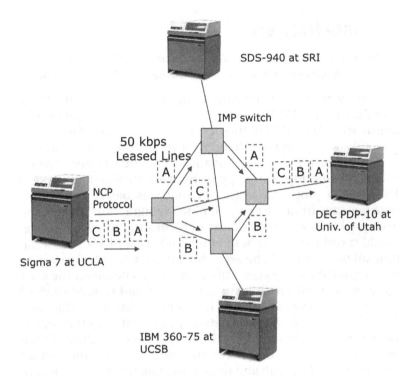

Figure 1.7 Schematic of ARPANET.

Each IMP would take data message of up to 8000 bits, shown as ABC in Fig. 1.7, from the computer connected to it; subdivide the data message into packets of 1000 bits shown as A, B, and C in the figure; add a header specifying destination and source addresses to each packet; and then based on the routing table, send the packets over which ever line was free to the next IMP toward the destination. The receiving IMP would acknowledge the receipt and repeat the process until the packet reached the destination. The destination would reassemble the packets in correct order to recover the transmitted data message, ABC. The success of ARPANET proved the viability of packet switching for data networks.

Soon ARPANET grew from 4 nodes to 23 nodes by 1971, 37 nodes by 1972, 62 nodes by 1974, and 111 nodes by 1977. The data transfer protocol on ARPANET was first Network Control Program (NCP) but was replaced by transmission control protocol/Internet protocol (TCP/IP) in 1984. This TCP/IP protocol was developed earlier by Vinton Cerf and Robert Kahn[10] in 1974.

Although ARPANET demonstrated the viability of packet switching over long distances, it was a private network built to demonstrate the concept of packet switching and to serve a very restricted set of users. With the growing use of computers, there was a growing commercial need to connect an organization's computers. The emphasis was first on connecting these computers in a building or in a campus in close proximity. The reason was due to the fact that at that point in time growth of computers was limited to large organizations and enterprises having computers arranged only in a local environment. Robert Metcalfe and David Boggs[11] developed Ethernet protocol in 1976 for computers in close proximity, for example, within a building which came to be known as local area network (LAN). By late 1980s, Ethernet-based LANs were common place. It would, however, take another about 25 years before Ethernet protocol would be extended to cover long distances, and with that growth now the importance of peering of such Ethernet-based networks, belonging to different carriers, is becoming critical. And that is the subject of this book. More on that is in later chapters.

As the popularity of computers was growing and enterprises were becoming global, there was now a need to connect these LANs located not only across the United States but around the globe. Only telecommunications companies with their PSTN networks had the capability to provide leased lines for such wide area networking or in short WANs. At first, it was based on 56 kbps leased lines, but as traffic grew and TDM-based multiplexing technologies arrived, these were replaced with T1 lines. This is shown in Fig. 1.8.

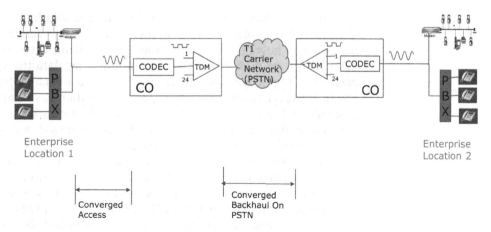

Enterprise Location 1

Enterprise Location 2

Converged Access

Converged Backhaul On PSTN

Figure 1.8 Schematic of TDM-based PSTN for connecting Enterprise LANs.

As the traffic grew even more, the T1 lines were replaced with 45 Mbps T3 lines. Both T1 and T3 are based on plesiochronous digital hierarchy (PDH) technology, and as the traffic grew further and as optical transmission capabilities became available by more advanced TDM technologies like synchronous optical network (SONET) and synchronous digital hierarchy (SDH). These provide higher performance and bandwidths.

Fig. 1.9 shows the mechanism of TDM and its use for data transport. As explained earlier, multiplexing involves combining and transporting many slow channels on a single faster channel. This required sizing the bandwidth of the TDM link to account for (1) the maximum traffic rate of the data stream; (2) encapsulating data frames, mostly Ethernet frames, coming from LANs in to payloads for the TDM network; (3) routing the payload; and finally (4) decoding the TDM payload to reconstruct data stream or Ethernet frames at the destination. Because data streams are bursty or sporadic in nature, the maximum bandwidth is always higher than the nominal bandwidth; the TDM links for data transmission over WANs are generally underutilized.

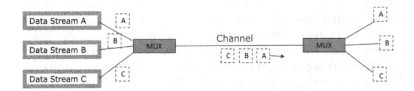

Figure 1.9 Time-division multiplexing.

Since TDM bandwidth is the resource that is sold by the carriers to the customers and customers pay for this bandwidth regardless of how much of traffic flows on that bandwidth, use of TDM links over WAN is not the ideal solution for data transport because of high costs and suboptimal use of the allocated bandwidth. However, from commercial perspective that was what was available in the 1970s and 1980s in the public network arena, it was widely deployed, well understood, and managed, and so it became the choice of many enterprises for data transport.

Meanwhile on the packet switched data network front, National Science Foundation (NSF) in 1981 created another network called CSNET to connect institutions not covered by ARPANET and in 1983 interconnected CSNET to ARPANET. This extended ARPANET too became congested, and so NSF built a backbone called NSFNET to which various networks could connect using TCP/IP protocol. This *inter*connected *net*works with NSFNET as backbone with attached regional networks came to be known as the *Internet*. This growth received tremendous boost when AT&T's Bell Labs made Unix operating system open in 1975, and University of California, Berkley, added TCP/IP protocol to the Unix operating systems and made it publicly available in 1983 as Unix BSD. In 1984 domain name service, developed by Paul Mockapetris,[12] added further boost to data networks. By 1989 the number of nodes on Internet reached 100,000, and so the backbone was upgraded to 1.544 Mbps (T1) from 56 kbps. In 1992 Internet Society was formed to provide leadership to this growing Internet, and in 1993 the backbone was upgraded from T1 to T3 (45 Mbps). It had taken 20 years for the Internet backbone to be upgraded from 56 kbps to 1.544 Mbps but took only 4 years to be upgraded from 1.544 Mbps to 45 Mbps! Personal computers (PCs) arrived in 1980s popularized by IBM-compatible PCs using first Microsoft's DOS and then Windows operating system on one hand and Apple with its (graphical user interface) GUI-based OS X operating system on the other hand. Contribution to the growth in Internet traffic from these PCs got a real boost when Tim Berners-Lee developed and his employer CERN provided the World Wide Web software[13] in public domain on April 30, 1993. Fig. 1.10 shows the evolution of the data communications.

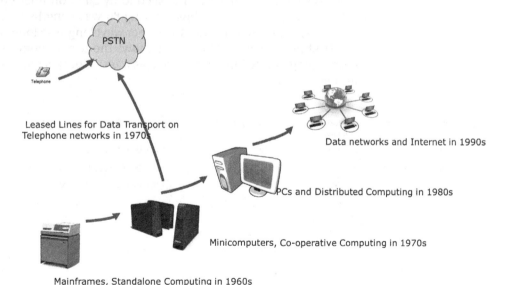

Figure 1.10 Evolution of data communication.

As you may have noticed from the examples of AT&T's Bell Labs making UNIX operating system open followed by University of California, Berkley, making UNIX BSD operating system publicly available and then CERN offering Web browser software in public domain, the growth of data networks was based on a new phenomenon of making software open to public called Open Source, and over time, many other important softwares were added to this category. It also led NSF to involve commercial Internet providers and to organizing the Internet as a combination of local, regional, and national Internet service providers (ISPs) and content providers. More often than not a company would play in all these areas simultaneously. On April 30, 1995, NSFNET was officially shutdown but NSF's new ISP-based architecture became the foundation for data networks popularly called Internet. We will discuss this architecture in more detail in Chapter 2.

Although the original concept of portable computers was embodied in Zilog Z80 released in 1981 by Adam Osborne, it did not catch on till 1999. Now, laptops are everywhere, and tablets, a form of laptops with touch screen and without attached keyboard, are flooding the markets. All these devices and much more including modern high-definition and ultrahigh-definition digital TVs are adding traffic throughput to Internet, which incidentally, as we will see in more detail in Chapter 8, is being called *Internet of Things*, at a pace that is in terabits per second (Tbps) and expected to reach 1.4 petabits per second by 2019-on fiber optic lines with bandwidths of 8 Tbps carrying 80 wavelengths [at the rate of 100 gigabits per second (Gbps) per wavelength]-a long way from 56 kbps copper lines! Fig. 1.11 shows the terminology used in describing bandwidth and storage as a result of this growth in Internet.

10^n	Prefix	Symbol	Short scale	Long scale	Decimal equivalent
10^{24}	yotta	Y	Septillion	Quadrillion	1 000 000 000 000 000 000 000 000
10^{21}	zetta	Z	Sextillion	Trilliard (thousand trillion)	1 000 000 000 000 000 000 000
10^{18}	exa	E	Quintillion	Trillion	1 000 000 000 000 000 000
10^{15}	peta	P	Quadrillion	Billiard (thousand billion)	1 000 000 000 000 000
10^{12}	tera	T	Trillion	Billion	1 000 000 000 000
10^9	giga	G	Billion	Milliard (thousand million)	1 000 000 000
10^6	mega	M		Million	1 000 000
10^3	kilo	k		Thousand	1 000
10^2	hecto	h		Hundred	100
10^1	Deca, deka	da		Ten	10

Figure 1.11 Terminology used in describing bandwidth of data networks.

The new portable devices are leading to increased mobility with rich audio, data, and video applications which has changed our world in a way that touches everyone. And talking about video applications, it is perhaps a good segue to our section on video or HFC networks.

1.3 Hybrid Fiber-Coaxial Networks

Television won't last because people will soon get tired of staring at a plywood box every night.
Remarks by Darryl Zanuck, movie producer, 20th Century Fox, 1946

Television popularly known as TV was invented in the 1930s, and first cable TV was available in the United States in 1948. Its purpose was to provide better TV signal to those whose reception was poor because tall mountains and buildings blocked over-the-air TV signals. Originally, cable plant, a term for cable infrastructure used in cable TV industry, was designed for one-way communications from head end (HE) to subscribers using the radio frequency (RF) spectrum of 5–500 MHz. This overlaps with the part of spectrum (110–130 MHz) over which air traffic controllers communicate with airplanes. This reuse of spectrum is possible because cable TV signals are sent over shielded coaxial cables which prevent any leakage of signals and thus avoid interference with the air traffic control communication. The frequency ranges from 5 to 50 MHz was left reserved for any signal from subscriber to network, but since in a one-way communication, it was not possible, so this low end of the spectrum was left unused. The actual TV signals, which were analog in nature started from 50 MHz going up to 500 MHz. Each channel occupied 6 MHz, and as a result, about 80 channels could be accommodated. Now this spectrum has been increased up to 1000 MHz, and so about 150 TV channels can be accommodated when signaling is analog.

Today, cable passes over 110 million homes in the United States, of these over 70 million home are subscribers. Cable TV companies also known as multiple system operators (MSOs) recognized that they have an important asset of cable that passed millions of households which could be used for high-speed Internet connection and voice communications. Their consortium released a Data over Cable System Interface Specification (DOCSIS) in March 1997 to transport IP data traffic over cable. This traffic, however, required two-way communication, and so the cable companies upgraded their cable plant with fiber optic cables to accommodate digital DOCSIS channel, and later analog video channels also became digital. However, many HFC plants still have analog video channels National Television System Committee (NTSC) even today. This new system is referred to as HFC which stands for Hybrid-Fiber-Coaxial system. In HFC, fiber optics cable runs up to a point, and then coaxial cable runs down the street also the drop from street to home, and cabling inside the home is coaxial cable. DOCSIS specification made use of the till then unused 5–50 MHz frequency range for upstream data communication actually 5–6 MHz is not used due to noise and 6–42 MHz is used for upstream data communication and 42–50 MHz is reserved for control purposes. Since then DOCSIS specification has been revised many times to accommodate (1) quality of service (QoS) needed for IP Telephony, (2) support for IP version 6, (3) enhance transmission bandwidth to 10 Gbps downstream and 1 Gbps upstream using 4096 quadrature amplitude modulation, and (4) to do away with 6 MHz–wide channel spacing and instead using smaller 20 to 50 kHz wide channel spacing based on orthogonal frequency-division multiplexing (OFDM) subcarriers. The latest version is DOCSIS 3.1. Fig. 1.12 shows a typical HFC-based triple-play service consisting of audio, video, and data (high-speed Internet access) services offered to subscribers by cable companies.

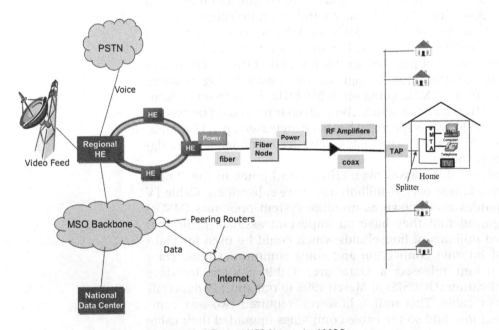

Figure 1.12 A Typical HFC Network of MSOs.

Just like the telephone companies have CO, MSOs have HE. Each HE serves about 20,000–40,000 homes through a fiber ring of fiber nodes in a metropolitan area network (MAN). Each fiber node has one to six coaxial cables attached to it, and each co-axial cable serves about 500–1000 homes. The access to each home is provided from a tap in the coaxial cable that has a splitter. One branch connects to the TV, and the other connects to an embedded multimedia terminal adapter (eMTA) commonly but mistakenly known as cable modem (CM). The CM in turn connects to a PC either directly or to a Wi-Fi router that allows a LAN inside the customer's premises. A phone is connected to eMTA by an RJ11 connector. The CM in the customer premises is terminated by a cable modem terminating system (CMTS) in the HE. HEs in turn are part of a regional area network (RAN) connecting to a regional HE or super HE. A typical super HE serves about 200,000–400,000 homes. These super HEs are connected to MSO backbone which through peering routers is connected to Internet to form wide area network (WAN). The MSO's backbone also connects to MSO's national data center where billing and other centralized information technology (IT) systems are housed.

In addition to the HFC infrastructure, cable companies also needed a set of IT applications to handle customer qualifications, provisioning, activation, service assurance, and billing for this triple-play IP service because their earlier system only handled cable TV or video services. This IT system is also known as OSS/BSS system and stands for operations support system/ business support system. An important capability of this OSS/ BSS system is to handle move, add, change, and disconnect the service popularly known in short as MACDs. In addition, the customer premises device that supports multiple services like data, voice, and video services may go through preprovisioned and provisioned states, and after the subscriber is active, each of the services may go through suspend or abuse state. For example, if a subscriber is found to abuse data services, then that service can be put under suspension due to abuse, but the service provider is legally obliged to keep limited voice service so that the subscriber can make emergency calls to 911. As a result of all these permutations and combinations, a subscriber and subscriber's services and the associated CPE go through many states. Kangovi[14] developed a state machine called service-linked multistate system (SLIMS) to handle all these multiple states. This state machine plays an important role in the OSS/BSS system. More on the OSS/BSS systems in Chapter 7.

Earlier, cable company's MAN and RAN networks used SONET, but they are now moving to gigabit Ethernet networks. This is because the cable companies are also moving to supporting all digital services while reducing TCO and consumers are increasingly using high- and ultrahigh-definition digital smart TVs requiring high bandwidth, low latency, and jitter.

This move to gigabit Ethernet networks is also helpful in supporting Internet Protocol television (IPTV) and Internet television both these are becoming very popular. IPTV frees up bandwidth because content remains in the network, and only the content selected by a customer is sent into customer's home, whereas in a typical HFC-based broadcast video technology, all the content constantly flows downstream to each customer's set-top box which first filters the unsubscribed channels, allowing only the subscribed channels and then handles switching from one channel to another, thus limiting the number of channels to the capacity of the HFC pipe to home. For this reason, IPTV is especially attractive where access to home is copper wire pair or satellite based. Internet television is also known as over-the-top (OTT) content because it goes over the top of the set-top box, in other words by-passes the set-top box. Set-top box is needed by both the HFC- and IPTV-based video transmission, whereas OTT content can be accessed directly by a smart TV connected to Internet. Examples of OTT services are Netflix, Hulu, and YouTube. Detailed discussions of HFC-based video technology, IPTV, and Internet TV (OTT) are beyond the scope of this book.

In this section, we examined how RF is being used by cable industry in their HFC networks to offer video, data, and voice services. However, the biggest consumers of the spectrum are the mobile telephone service operators as we will see in the next section.

1.4 Wireless Networks

Cellular phone will absolutely not replace local wire systems
Remarks by Martin Cooper—Director of Research and Developer of First Handheld
Cell Phone, in 1981.

The purpose of this section is to give a high-level overview of the evolution of wireless communication to show its impact on access and backhaul portions of the networks and to show the linkage with carrier Ethernet networks (CENs) and peering of CENs. This section is not a detailed reference on wireless communication or wireless networks. Therefore exhaustive coverage is out of scope here. The field is rapidly changing, and there are many excellent references on this subject for those interested in details. It is also important to note that the term wireless communication, though commonly used in connection with mobile phone service, is a general term for referring to unbounded transmission technologies. This field is divided into categories based on frequency. Examples in the low end of the frequency of the electromagnetic spectrum are radio, terrestrial microwave, and satellite microwave links and at the high end of the frequency are laser and infrared links. We know from Shannon's information theory[6] that higher the frequency more is the bandwidth but higher frequency also requires more power, for example, Wi-Fi routers inside homes or buildings use frequency range of 2.4–5 GHz, but they are connected to a power outlets and do not use batteries. On the other hand, mobile phones use radio link in a relatively lower frequency range to conserve battery power. With improved design, this frequency is increasing as we will see later in this section. The terrestrial microwave links in the low GHz range are mostly used as line-of-sight beam for a point-to-point connection for WANs. On the other hand, higher frequency satellite links are mostly used for large footprint coverage in areas which are remote and cannot be reached otherwise. There are many companies offering satellite-based TV services; however, satellite communication is not particularly suitable at this point for common high-speed data and near real-time voice services. The reason for this is because the microwave signal has to travel from point of origination on earth to reach geosynchronous satellites in an orbit at a distance of 22,500 miles and then return to earth to its destination. That adds up to 45,000 miles and takes up to 0.25 s. This latency is an order of magnitude higher compared to latencies of 0.005–0.05 s normally seen in voice and data transmissions. Now, coming to high-frequency laser links, these are used for short distances because these are not very effective when unbounded as they are disturbed by atmosphere. Laser, however, is extensively used in bounded transmissions in conjunction with fiber optic cables. As far as the high frequency infrared links are concerned, these are also used over short distances for high bandwidth in unbounded transmissions for example in PC-to-PC connections and wireless LANs. Now that we have addressed some preliminaries about wireless communication, our focus in the rest of this section will be on the application of wireless communication for mobile phones or smartphones.

It was James Clerk Maxwell[15] who showed by mathematical formulations in 1864 that electromagnetic waves can travel through free space. In 1888, Heinrich Rudolf Hertz[16] proved by his experiments the accuracy of Maxwell's theory. Initially, these waves were known as Hertzian waves and after about 20 years were termed as "radio" derived from radiation of electromagnetic waves. Their frequency in cycles per second is called hertz in honor of Heinrich Rudolf Hertz. It took another 6 years when in 1894, Guglielmo Marconi[17] built the first commercially viable system for radio transmission ushering in the era of radio communications.

It, however, took another 50 years for the intersection of telephony and radio communication in 1947 when old AT&T introduced mobile telephone system. Because of the availability of only three channels and high costs, it did not catch on. In the 1960s, a new system launched by Bell Systems, called improved mobile telephone service (IMTS), brought many improvements like direct dialing and higher bandwidth. The first analog cellular systems were based on IMTS and developed in the late 1960s and early 1970s. The systems were "cellular" because coverage areas were split into smaller areas or "cells," each of which is served by a low power transmitter and receiver. Then in 1978 old AT&T's Bell Labs developed an advanced mobile phone system which gave higher capacity. It was used through 1980s in to 2000s before transitioning to digital system. These systems for mobile communications saw two key improvements during the 1970s: the invention of the microprocessor and the digitization of the control link between the mobile phone and the cell site albeit voice transport was still based on an analog system. The mobile phones and associated cellular networks based on analog narrowband systems were called 1G and transmitted only voice. The phones were bulky and on the back end connected to PSTN.

Second generation (2G) digital cellular systems specifications were first developed at the end of the 1980s. These systems digitized not only the control link but also the voice signal. They allowed limited short message service (SMS) or texting capabilities. The new system provided better quality and higher capacity at lower cost to consumers. They were based on global system for mobile communication (GSM) technology used by some and code-division multiple access (CDMA) technology by others. It was based on 900-MHz central frequency with a bandwidth of 25 MHz divided into 124 tracks. The GSM architecture is shown in Fig. 1.13.

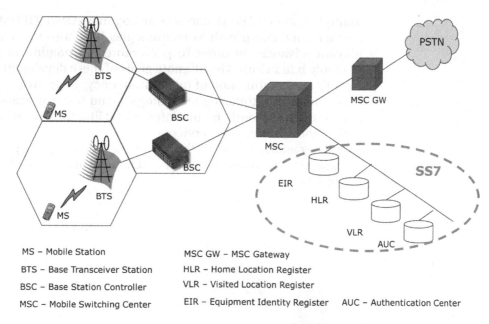

MS – Mobile Station

BTS – Base Transceiver Station

BSC – Base Station Controller

MSC – Mobile Switching Center

MSC GW – MSC Gateway

HLR – Home Location Register

VLR – Visited Location Register

EIR – Equipment Identity Register   AUC – Authentication Center

Figure 1.13 GSM architecture.

Mobile operators use radio spectrum to provide service. Although spectrum is created by God, it is allocated by the World Radio Conference (WRC) and auctioned by governments to service providers who turn it into bandwidth governed by Shannon's information theory[6] which, of course, is the gospel in the field of communications. Therefore, it all fits together. But it is a scarce resource, shared, and used by many industries. At WRC1993 spectrum allocation for 2G was agreed upon. Cellular networks are made of cells, normally in group of seven. Each cell has a base station called base transceiver site (BTS). This has antennae, transmitter and receiver, and switching equipment. The size of the cell depends on population density. BTSs are connected to Mobile Switching Office/Center (MSO/ MSC) through base station controller (BSC) either by fixed line or by microwave called backhaul. MSO handles authentication and SS7 signaling and in turn is connected to PSTN. One of the important functions of cellular technology is to manage cell handoff in a seamless fashion as the users travel around. Since spectrum is a scare resource, various, progressively improved, multiplexing mechanisms have been developed over time including frequency-division multiple access (FDMA), time-division multiple access (TDMA), code division multiple access (CDMA), and now orthogonal FDMA (OFDMA). Furthermore, compressions technologies have also made significant advances in order to pack more information in the available bandwidth. The ultimate goals of these developments were to use combination of higher frequency range, frequency reuse, better multiplexing technologies, and faster processing to increase bandwidth, reduce latency and jitter at a lower cost, and provide multimedia services.

2.5G technology–based cellular networks, such as global packet radio service, provided an overlay of data network. This is shown in Fig. 1.14.

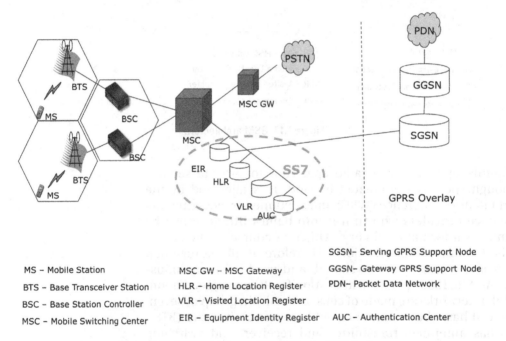

MS – Mobile Station	MSC GW – MSC Gateway	SGSN– Serving GPRS Support Node
BTS – Base Transceiver Station	HLR – Home Location Register	GGSN– Gateway GPRS Support Node
BSC – Base Station Controller	VLR – Visited Location Register	PDN– Packet Data Network
MSC – Mobile Switching Center	EIR – Equipment Identity Register	AUC – Authentication Center

Figure 1.14 GPRS Overlay on GSM system architecture.

At WRC2000, the spectrum allocation for third generation (3G) was agreed upon. This allocation uses spectrum from 2G and extended it to 3 GHz. 3G systems provided faster communications services, including voice, fax, and Internet, anytime and anywhere with seamless global roaming. The first 3G networks were deployed in Korea and Japan in 2000 and 2001. These 3G networks required enhancing GSM to Universal Mobile Telecommunications System (UMTS) and CDMA to W-CDMA (wideband code division multiple access). Global standard for 3G is based on ITU's International Mobile Telecommunications(IMT)-2000 standard and has opened the way to innovative applications and services like multimedia entertainment, information, and location-based services. The high-level architecture of 3G wireless system is shown in Fig. 1.15.

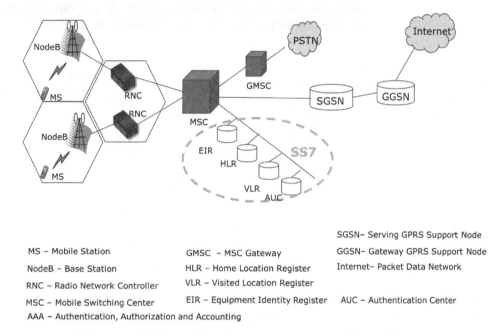

MS – Mobile Station

NodeB – Base Station

RNC – Radio Network Controller

MSC – Mobile Switching Center

AAA – Authentication, Authorization and Accounting

GMSC – MSC Gateway

HLR – Home Location Register

VLR – Visited Location Register

EIR – Equipment Identity Register

SGSN– Serving GPRS Support Node

GGSN– Gateway GPRS Support Node

Internet– Packet Data Network

AUC – Authentication Center

Figure 1.15 Third-generation (3G) system architecture.

Long-term evolution (LTE)/advanced LTE is the new wireless technology that uses IP-based system optimized for data delivery. In this technology, there is no separate core for data and voice. In order to support traditional requirements for (1) signaling, (2) standard digitized voice format for translation between standard telephony format and VoIP format, and (3) QoS on a packet switched network, the industry body called Third-Generation Partnership Project recommends implementing IP Multimedia Subsystem (IMS) for handling voice communication on LTE networks. LTE has central frequency around 2.5 GHz with bands of 1, 3, 5, 7, 8, 11, 13, 17, 25, 26, 40, or 45 MHz. Fig. 1.16 shows the architecture of LTE/advanced LTE system. LTE system uses an Evolved Node B (eNodeB) which is essentially a base station; a Mobility Management Entity (MME); a home subscriber server (HSS) which is a super home location register from earlier technology and combines the Authentication, Authorization, and Accounting functions; a serving gateway (S-GW); and a packet data network gateway (P-GW). The network of eNodeB is called evolved universal terrestrial radio access network (eUTRAN), and combination of MME, S-GW and P-GW is called evolved packet core (EPC). The MME and HSS handle all functions related to subscriber access to the network including authentication, authorization, accounting, roaming rules, etc. The S-GW passes data between subscriber and network, and P-GW provides connection to external data network, that is, Internet.

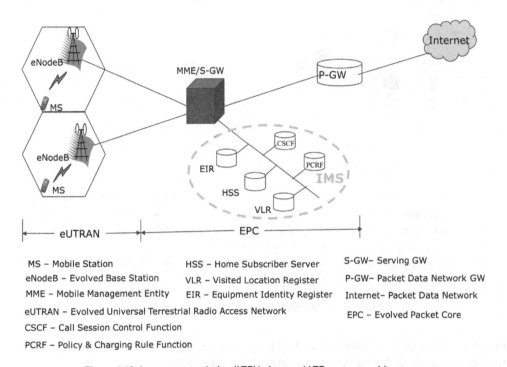

MS – Mobile Station
eNodeB – Evolved Base Station
MME – Mobile Management Entity
eUTRAN – Evolved Universal Terrestrial Radio Access Network
CSCF – Call Session Control Function
PCRF – Policy & Charging Rule Function

HSS – Home Subscriber Server
VLR – Visited Location Register
EIR – Equipment Identity Register

S-GW– Serving GW
P-GW– Packet Data Network GW
Internet– Packet Data Network
EPC – Evolved Packet Core

Figure 1.16 Long-term evolution (LTE)/advanced LTE system architecture.

LTE uses two different types of radio links: one for downlink, that is, from tower to mobile device and another link for uplink, that is, mobile device to tower. By using different links, LTE optimizes battery life and network connection. For downlink, LTE uses OFDMA. OFDMA mandates multiple in, multiple out (MIMO). Having MIMO means a device has multiple connections to a single cell which reduces latency and improves stability of connection and increases throughput. For the uplink, LTE uses discreet Fourier transform spread OFDMA (DFTS-OFDMA) scheme for a single-carrier FDMA (SC-FDMA). This improves the peak-to-average power ratio for uplink. Since LTE uses IMS for voice on IP, it is called voice over LTE (VoLTE).

Initially, LTE was called 4G but LTE did not fully meet IMT-advanced specifications from ITU. Only LTE advanced meets the

IMT-advanced specification. However, now the IMT-advanced specification has been revised, and this revision is considered 4G. As a result, debates still continue if LTE is 4G and what exactly is 4G. LTE, however, has opened the way for data only network from the hybrid voice–data networks, and it has increased bandwidth to a point where it will be in competition with digital subscriber line (DSL) and cable for local access. It also has opened the way for improving battery life and advanced radio links.

The evolution of wireless communication from 1G to 4G is shown in Fig. 1.17.

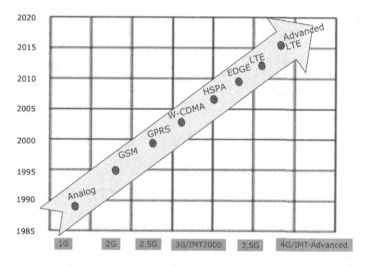

Figure 1.17 Evolution of wireless communication.

The growth in the mobile data traffic is shown in Fig. 1.18. It shows that in about 4 years, the mobile data traffic has increased about 20 folds or by about 2000%!

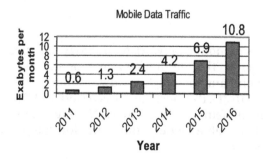

Figure 1.18 Global mobile data traffic growth.

Since mobile phones and the associated cellular networks represent a huge market, today this field is at the intersection of market forces, technological changes, turf battles, politics, and protecting investments on the installed bases. But the trend is clear-it is for higher bandwidth, low latency, low jitter, transition of voice to data, and support of multimedia services at low cost. Fig. 1.19 shows the impact of mobile technology on the evolution of data communication. The figure is somewhat complex because the current situation is complex.

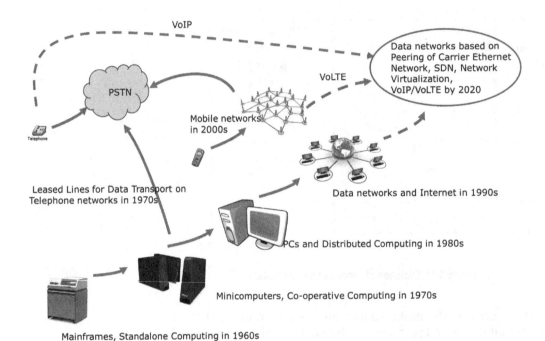

Figure 1.19 Impact of mobile technology on the evolution of data communication.

Regardless of mobile network technology, it is clear from the above discussions that mobile service depends on radio access on the front end and on a high bandwidth network on the back end. From 1G to 3G, this backhaul was based on TDM networks, but now this is being replaced by CENs (Fig. 1.20). This gives continuously variable higher bandwidth, low latency, and jitter and reduces TCO.

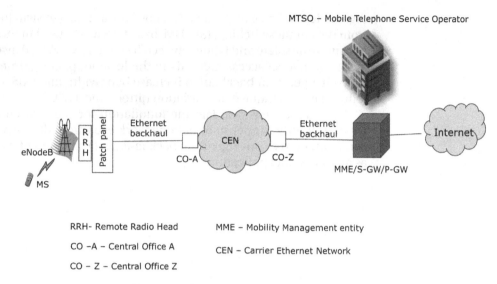

Figure 1.20 Access and backhaul in mobile network.

And that brings us to our next chapter where we will peal the onion little further insofar understanding the diversity of local access and convergence in the backhaul is concerned.

1.5 Chapter Summary

In this chapter, we surveyed the landscape vis-à-vis PSTNs, data networks, HFC networks of cable operators, and wireless networks. This survey identified the origins of many important building blocks developed especially in connection with PSTN. These building blocks included organization of network into access loop and trunk, switching and multiplexing technologies, Nyquist theorem for sampling rate to digitize analog signals including voice, and Shannon's information theory. The chapter described how these building blocks proved to be important for all other types of networks as well including data networks. Similarly, an important concept was proposed during the development of data networks when NSF defined a new architecture for Internet based on national, regional, and local ISPs. This led to organizing the data networks in to LAN, MAN, RAN, and WAN.

This overview chapter also interleaved the technological developments with the changing legal and business landscapes and helped in identifying some important trends including (1) digitization of voice and video, (2) move toward packet switching from the circuit switching, (3) TDM-based networks used in PSTN becoming obsolete and being replaced by data networks, (4) availability of diverse access methods in the local loop, and (5) need for convergence in backhaul to increase bandwidth and QoS and reduce latency, frame delay variation (jitter), and TCO.

This chapter has also laid the foundation for the next chapter to drill little deeper in to data network evolution, diversity of access methods, convergence of backhaul, and transition of voice as another data application.

2

SHAPING OF DATA NETWORKS

We shape our buildings, thereafter they shape us.

Remarks by Winston Churchill

Survey of different types of networks in Chapter 1 showed that the Internet, which is the network of data networks, is the most dominant network today. Rapid and extensive growth of the Internet is mainly due to the fact that it enables sharing of information without boundaries. This growth is supported, in large part, by the sound architecture that was laid out during its evolution. This evolution, however, was not bereft of controversies, issues, differences of opinions, and multitude of approaches. But from this chaos emerged some winners, strengthened by intense scrutiny and debates.

This chapter provides an overview of data network evolution including different methods of packet switching, OSI seven layer model, various data networking protocols, and NSF's architecture based on NAPs and ISPs.

This chapter also covers convergence in backhaul which is good for operators because it reduces TCO while increasing bandwidth, lowering latency, and jitter, thus enabling the operators to support voice, video, and data services on the same backhaul. This convergence in backhaul is coupled with diversity of choices in access loop, some based on installed base of copper wire-pair from PSTN and co-axial cable from HFC networks, others based on newer technologies like fiber optics and radio links. Description of different types of access loop is also included in this chapter. This rich diversity of choices in access loop is instrumental in the rapid growth of customer base while giving choice of a service provider to customers which in turn is keeping prices competitive. Finally, this chapter examines a typical approach that is enabling voice to become another data application, thus making PSTN obsolete and moving all applications on to data networks.

Peering Carrier Ethernet Networks. http://dx.doi.org/10.1016/B978-0-12-805319-5.00002-2

2.1 Protocols, Models, and Architecture of Data Networks

ARPANET in 1969 demonstrated the feasibility of packet switching ushering the era of data networks. It showed that cost of dynamic allocation for packet switching would be lower than the cost of preallocation used in circuit switching by PSTN. This economic advantage led to the push for data networks.

ARPANET had used IMPs (Fig. 1.7 in Chapter 1) which used message-based packet switching. IMPs lacked a flow control mechanism. When the outflow of messages from host computer exceeded IMP's capability to process them, the only mechanism that was available was to shutdown the flow of messages from host computer. This shutdown not only prevented outflow from but also inflow of messages to host computer. This of course was not a good solution.

In 1973, Louis Pouzin[18] in France developed the "datagram"-based packet switching. Host computers were connected to a network called Cyclades which included both hosts and subnetworks. It was the responsibility of the sending host computers to generate or assemble datagrams and sequence numbering and flow controlling them. The subnetworks transmitted these datagrams in whatever order they arrived and delivered on different available paths to receiving host computer which then assembled datagrams in correct sequence to recover the data message. Thus bulk of tasks related to creating and assembling datagrams and flow control were done by host computers and not network nodes.

Around the same time, Remi Despres[19] also in France developed another approach to packet switching called "virtual connection." This network was called RCP, and here the flow control was done by network nodes and not by host computers. This meant that host computers were not required to make substantial modifications to handle virtual connections and that was a strong marketing advantage of this approach. Table 2.1 gives a comparison of these two methods of packet switching.

Table 2.1 Comparison of Datagram and Virtual Connection–Based Packet Switching

Item	Datagram	Virtual Connection
Connection setup	Not needed	Required
Addressing	Each packet contains full destination address	Each packet contains a short VC identifier

Table 2.1 Comparison of Datagram and Virtual Connection–Based Packet Switching—continued

Item	Datagram	Virtual Connection
State information	Switch does not hold state information	Switch holds in a table VC state information
Routing	Each packet is forwarded independently	All packets follow same route
Effect of switch failures	None, except for packets lost during crash	All VCs passing through a failed switch are broken
QoS support	Difficult	Easy
Congestion control	Difficult	Easy

Almost in parallel to the work on datagram and virtual connections (VCs), efforts were underway in many countries between 1973 and 1976 to build public data networks within few years after the APRPANET demonstrated the benefits of packet switching. For example, TELENET[20] in the United States, EPSS[21] in the United Kingdom, TRANSPAC[22] in France, DATAPAC[23] in Canada, and DX-2/NTT[24] in Japan represented efforts to build public data networks. It was soon recognized that to connect these public data networks, there was a need for a standard otherwise host computers in different countries would not be able to network with each other. To this end, the Consultative Committee for International Telephony and Telegraphy (CCITT) Recommendation X.25 protocol was developed and agreed upon in March 1976. This protocol provided for 4095 VCs on a single full-duplex leased line to data network and included call setup and disconnect procedures. Each VC had a mechanism for flow control as well. By 1978, all five major public data networks implemented X.25, and in addition, TELENET in the United States and DATAPAC in Canada were interconnected, demonstrating the advantage of adopting a data networking standard.

X.25 was a major step forward but it had some deficiencies including reliability. If a VC failed during a call, there was no reconnect facility. Datagram-based packet switching on the other hand was more reliable, if there were multiple paths between sending and receiving hosts, then datagrams could take one or the other route. Also, VCs had costs associated with memory and call setup. Datagrams, however, had higher overheads because each datagram of 128 bytes had 25 bytes of header compared to only 8 bytes for X.25. Many improvements including X.28, X.29, and X.75 appeared within 2 years after the release of X.25.

Although these efforts were continuing to build data networks and interconnect them to send data over wide areas, there was also, in parallel, a growing interest to network computers in a local area, say within a building or a campus. To achieve that goal, a protocol called Ethernet[11] was proposed in 1976. This protocol was adopted in 1983 by IEEE, with some minor changes, as its IEEE standard 802.3.[25] Ethernet will be covered in more detail in our next chapter.

There was also work in progress to develop a standard architecture model to describe the data networking systems. In 1983, the efforts of ISO and CCITT (its name since then has changed to ITU-T) were merged to arrive at the Basic Reference Model for Open Systems Interconnections in short OSI model. In 1984, this was published by ISO as standard 7498 and by ITU-T as X.200.[26] This was a major development and has guided all the work on computers and data networks since then. There are extensive references on this topic, and it will be covered in some detail in this chapter also because of its relevance to subsequent discussions. OSI model was based on learnings from ARPNET, CYCLADES, and other data networks. This model is based on dividing a data communication system into seven layers and associating each layer with a set of specific protocols. Each layer provides service to a layer above it and calls the services from the layer below it. The protocols associated with each layer enable a host to interact at the same layer with another host. Fig. 2.1 shows the OSI seven layer model along with examples of associated protocols

Number	Layer	Unit of Message	Protocol Example
7	Application	Data	HTML Class
6	Presentation	Data	HTML
5	Session	Data	HTTP
4	Transport	Segment	TCP
3	Network	Packet	IP
2	Data Link	Frame	Ethernet
1	Physical	Bits	

Figure 2.1 OSI seven layer model.

and the unit of data message with each layer. Layer 1 known as the physical layer is responsible for electrical or optical interface to the communications medium such as copper wire-pair, optical fiber, or radio link. Its services are used by layer 2 which is called the data link layer. This layer provides the service to layer 3 and uses the services of layer 1. Data link layer is responsible for taking the packets from layer 3 and framing them and transmitting frames to layer 1. Layers 1 and 2 make up subnetwork in the sense they are local in nature and confined to a segment of the network.

Layer 3 known as the network layer is responsible for data transfer across the network. It gets data segments from layer 4 and converts them to packets to transmit them to layer 2. Layer 4 known as the transport layer is responsible for the functional and procedural means of transferring variable-length data sequences from a source to a destination host via one or more networks, while maintaining the quality of service functions. The transmission control protocol (TCP) and the user datagram protocol (UDP) are examples of protocol at Transport layer. TCP provides end-to-end reliability of transmission of data, whereas UDP does not. Layers 3 and 4 concatenate different data networks. These four layers, namely physical, data link, network, and transport layers together are known as infrastructure of a data network. Layer 5 known as session layer is responsible for managing session and establishing and releasing connections. This layer also handles full-duplex, half-duplex, or simplex operation. Layer 6 known as presentation layer is responsible for converting bits into readable format. It is also responsible for encryption and decryption of data. Finally, the layer 7 known as application layer is responsible for managing the applications. This layer provides APIs to applications which in turn provide interface between the user and the communication system for capturing input from the user or providing output to the user. It is important to note that the application layer is not applications itself but provides services to applications. Layer 1 and 2 are more hardware based and layers 3 to 7 are software based.

Fig. 2.2 shows an example how the OSI seven layer model is applied to describe data communication between a sending host and a receiving host connected by data network consisting of fiber optic cables and switches. An application on the sending host calls the APIs to use the services of the application layer. This layer 7 captures the user data in natural language and adds a header and encapsulates it. Layer 7 then calls the services of the presentation layer which is the layer 6 to convert user data into binary code by using, for example, ASCII codes.

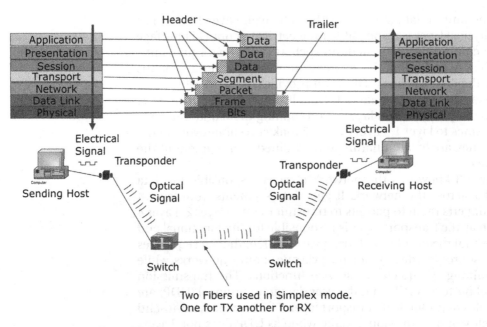

Figure 2.2 Schematic of an example of OSI seven layer model.

Presentation layer also handles bit ordering in "Big Endian" or "Little-Endian" method. This layer is also responsible for any encryption. Layer 6 then adds a header to this binary data and passes it to layer 5 which is the session layer. This layer 5 is responsible for establishing the connection with the remote host. Once the connection is established, this layer handles the data transfer in simplex, half-duplex, or full-duplex mode. At the end of data transfer, this layer is responsible for closing the connection. Finally, this layer passes the data with its own header to layer 4 which is the transport layer. Most common protocols implemented in this layer are TCP and UDP. TCP is connection based and UDP is connectionless. This layer is responsible for flow control, sequencing, and error recovery. This layer converts the large data sets received from upper layer into what is called segments which is the protocol data unit (PDU) of this layer. It then numbers each segment with a sequence number so that the receiving host's layer 4 can reassemble the segments in proper order. The error recovery function in TCP protocol is achieved by a positive acknowledgment from layer 4 of the receiving host to layer 4 of the sending host. If this is not received within a specified time, then the segment is resent. Layer 4 calls the services of network layer which is layer 3. This layer provides connection services by discovering route and adding header to packets which is the term for this layer's PDU. This header has the IP address of the destination based on route discovery by querying a DNS. Layer 3 packets are examined by routers to determine how to route the packets. The most common protocol implemented in this layer is the IP protocol. Layer 3 then passes packets to layer 2 which is the data link layer. Most common protocols implemented in this layer are Ethernet and WiFi. Today, majority of Internet traffic starts or ends on Ethernet. The PDU for this layer is called frame. This layer encapsulates packets coming from layer 3 into frames, it adds its header and trailer to establish link in the sub-network, manage error detection and flow control. Frames of layer 2 are examined by Ethernet switches to determine links for forwarding frames. This layer has two sub-layers called MAC and logical link control (LLC) sub-layers. We will cover this layer 2 in detail in the next chapter on Ethernet. This layer calls the services of the layer 1 which is the physical layer. Layer 1 provides the clocking, synchronization, and encoding services to convert the binary data coming from layer 2 into electrical signal and put that signal on the media. In case of transport using fiber optics, there is a transponder associated with this layer as shown in Fig. 2.2. Multiplexing, if needed, is handled by this layer.

After the development of OSI seven layer model, support for X.25 protocol was dropped in 1984 because it was covering layers 1, 2, and 3 together and other protocols conforming to OSI seven layer model provided better functionalities. X.25 also had an issue with recovering from link failure. Initial proposal for Frame Relay protocol was presented to the CCITT in 1984 but due to lack of interoperability and lack of complete standardization, frame relay did not experience significant deployment until 1990. Asynchronous transfer mode (ATM) protocol was developed in 1988 but was not commercialized until the mid-90s. Both were layer 2 protocols developed for data transport over wide areas and could also be used on layer 1 for limited distances and bandwidths but for larger bandwidths and distances, both frame relay and ATM used, for example, protocols like SONET/SDH at layer 1. The main issue was that all the three protocols X.25, ATM, and frame relay needed specific equipment at the customer premises as well as in networks. This increased TCO and training requirements for customers so that they could interface the customer premise equipment with their LAN and configure the equipment to convert their predominantly Ethernet frames to a data format suitable to one of these protocols. This in turn also increased processing time and hence latency. The other alternative was to take Ethernet frames from LAN traffic and move them over WAN using TDM circuit switching–based T1/T3/SONET/SDH networks with fixed bandwidth which was more expensive than VC–based connections that X.25, ATM, and frame relay offered. Neither option was ideal but those were the only options at that point in time. However, this situation changed around 2004 when Ethernet advanced to a point where it could be used over WAN in addition to LAN, where it was well entrenched. More on this in next chapter.

There was another important development around 1992 that was as important as the OSI seven layer model. This development was based on the blueprint proposed by NSF[27] for the Internet. It divided the Internet into national, regional, and local ISPs and provided for connecting networks of national ISPs at network access points (NAPs). This blueprint, shown in Fig. 2.3, had two important implications, first one was related to involving commercial providers in the growth of the Internet, and second was to provide a hierarchy of building blocks in organizing the Internet.

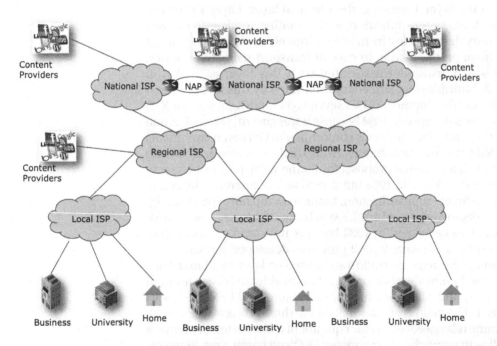

Figure 2.3 NSF architecture for Internet.

Around 1993, the national ISPs included AGIS, GTE, MCI, UUNET, PSINet, and Sprint. These were interconnected or peered at four network access points (NAPs) built in Chicago, Palo Alto, Washington, D.C., and Pennsauken. Two of the first regional ISPs were MERIT and BARRNet and some examples of the local ISPs in the United States included old A&T, MediaOne, TIAC, and

Ultranet. At that time, the content providers included Prodigy, CompuServe, and AOL. This picture soon changed with the Telecommunications Act of 1996, when many new ISPs joined the existing ISPs and ISPs started to play in all three roles of national, regional, and local ISPs. Also, additional Internet exchanges were built to supplement original four NAPs. These were called metropolitan area exchanges (MAEs). This peering of data networks at NAPs and MAEs constituted the Internet. These developments soon changed the ISP-centric architecture to a geography-centric architecture resulting in dividing the overall data network into LAN, MAN, RAN, and WAN. This is the topic of our next section.

2.2 Convergence in Backhaul

As data network was taking shape driven by OSI seven layer model and NSF architecture, the methods for accessing voice, video, and data services were also going through evolution. The situation that existed in early 1990s is depicted in Fig. 2.4.

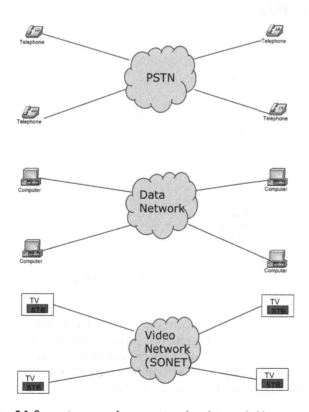

Figure 2.4 Separate access for separate voice, data, and video networks.

At that time, different accesses were needed for different services provided on different networks. Voice service was provided on PSTN which was accessed by telephones on copper wire pair. For computers to access data services, a data circuit terminating equipment (DCE) was needed to connect to data networks, and video services were accessed by TVs via set top boxes connected to video networks through coaxial cables. In short, there were three different networks providing three different services, and subscribers needed three different connecting devices.

This situation changed with the development of DSL technology and also the DOCSIS-based cable modem or eMTA technology in later half of 1990s. With DOCSIS, subscribers were able to get three different services using one access, and with DSL, they were able to get data and voice on one phone line; however, in the backend, there were still three different networks. This situation is shown in Fig. 2.5. Although it made it easier for subscribers, it was not simple for service providers. They had to maintain many different networks and that increased TCO.

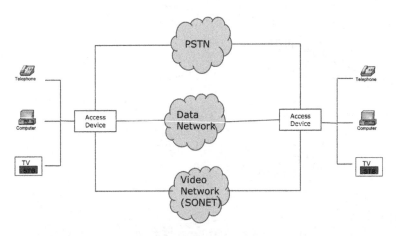

Figure 2.5 Multiple services on different networks with choice of access.

With data networks becoming the most dominant network with Ethernet becoming dominant protocol for layer 1 and 2 and TCP/IP for layer 3 and 4, and with growing expansion of VoIP and VoLTE for voice, stage has been set for true convergence of backend. This situation is depicted in Fig. 2.6.

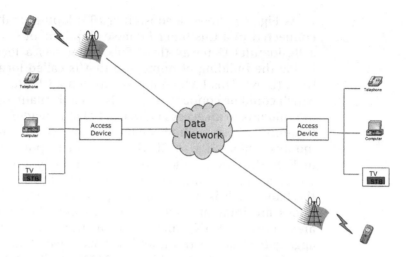

Figure 2.6 Multiple services on one network with choice of access.

This converged backend together with choice of access loop combined with the geography-centric architecture derived from the original NSF architecture for Internet is shown in perspective in Fig. 2.7.

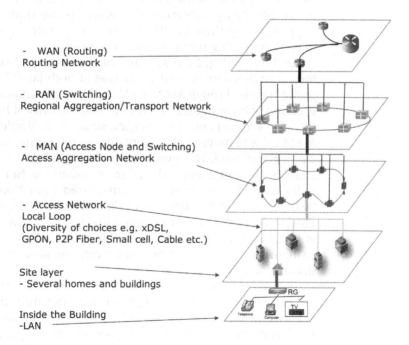

- WAN (Routing)
Routing Network

- RAN (Switching)
Regional Aggregation/Transport Network

- MAN (Access Node and Switching)
Access Aggregation Network

- Access Network
Local Loop
(Diversity of choices e.g. xDSL,
GPON, P2P Fiber, Small cell, Cable etc.)

Site layer
- Several homes and buildings

Inside the Building
-LAN

Figure 2.7 LAN, MAN, RAN, and WAN in a Perspective.

As Fig. 2.7 shows, a subscriber's TV, laptop, and phone are connected to a Customer Premises Equipment (CPE) such as a Residential Gateway (RG). Subscriber has a local network inside the building or home and that is called local area network (LAN). This LAN connects through a CPE to a local loop which could be copper wire pair, HFC cable plant, optical fiber, or radio link to an access node in a CO or a headend or a tower. Common access methods include various flavors of DSL commonly designated as xDSL, GPON, point-to-point optical fiber or DOCSIS-based cable service. Increasingly, one can access voice, video, and data services using radio link as well because the bandwidth is becoming comparable to xDSL. The access nodes are interconnected by a network called metropolitan area network (MAN). Important functions of this MAN are to aggregate all access traffic and switch at layer 2 (Ethernet layer) to a destination if reachable from MAN and if not, then to send the traffic to a layer called Regional Area Network (RAN). This network aggregates traffic from various MANs and it also does switching at layer 2 to direct traffic to its destination if reachable from here else to a wide area network (WAN). The main functions of WAN are to aggregate traffic from RANs and route the traffic by examining layer 3 (IP layer) packets. This process of "switch many, route once" speeds up the traffic by reducing processing time. WANs of different operators are interconnected at peering points to form the Internet.

As shown in Fig. 2.7, data enters or leaves subscriber's LAN, and format of this data is mostly Ethernet at both layer 1 and 2. This traffic travels through MAN, RAN, WAN, or the Internet again as Ethernet at layer 1 and 2. In other words, majority of Internet traffic today starts or ends on Ethernet. Similarly, in wireless network, the trend is to move to all IP networks with Ethernet at layer 1 and 2 in the backhaul. This evolution is shown in Fig. 2.8. In 3G, subscribers communicate with cell tower known as NodeB through radio link. This traffic is then transported from NodeB to RNC by backhaul which is based on TDM, ATM, and now increasingly Ethernet-based IP networks depending on the available transport at that location. This traffic from RNC is then directed to mobile switching center (MSC) where voice calls are sent over PSTN and data traffic is sent over the Internet.

In case of LTE, subscribers communicate with cell towers known as eNodeB (evolved NodeB), and from there, traffic is transported over switched Ethernet backhaul to a mobility management entity (MME) via a serving gateway (SGW) which handles the functionalities of RNC and MSC. MME handles the control functions of the SGSN of 3G networks. In other words in

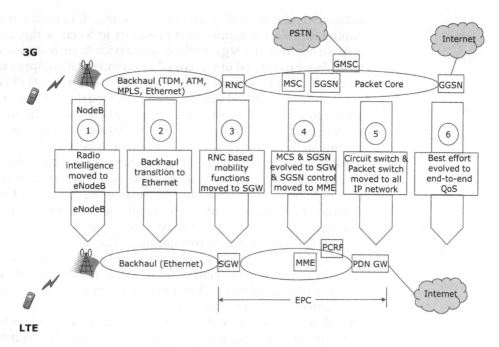

Figure 2.8 Backhaul convergence in wireless networks.

LTE, SGW and MME have been designed to take over functions distributed over RNC, MSC, and SGSN. Also, Fig. 2.8 shows that there is no connection to PSTN. That is because voice is now treated as another data application, and voice calls are now sent on the data network.

In view of these applications of Ethernet in backhaul of both wireline and wireless networks, Ethernet traffic is growing by leaps and bounds. This growth is leading to increasing interest in peering MAN and RAN networks of different operators at Ethernet layer and that is the topic of this book and will be covered in Chapter 5. But before that, in next section, we will cover the choices subscribers have for accessing voice, video, and data services. It is the access that provides the link between subscribers' LANs and the outside world of the Internet, and therefore it is important to know about access.

2.3 Diversity in Access

Owing to wide spread use of wireline-based telephone service in the past, copper wire pair is the most ubiquitous media installed at homes and office buildings around the world. Therefore it was natural for telecommunications operators to leverage this available access from the legacy telephone service to deliver new services. As we discussed in section 1.1 of Chapter 1, this copper wire pair is commonly known as unshielded twisted pair (UTP) copper wire. Initial method was based on providing data service using analog modems at the subscriber's end and ISP's end. The analog modem would convert data bits into analog signals and transmitted this analog signal on telephone network to ISP's modem where the signal was reconverted to data. It is important to note that the analog transmission was acoustic. The word modem is derived from "Modulator–Demodulator" because of its function of converting data signal to analog signal and back to data signal. Just as phones on either ends connected two human beings, modems on either ends connected two computers. The analog modem was not capable of supporting both voice and data services on the same line, and so subscribers had to either use data or voice at a time and that led to lot of inconvenience to subscribers. Many subscribers bought two telephone connections, one for voice and the other was used for data services. But having two phone lines was expensive. In addition, the bandwidth was low. This situation changed with the 1996 deregulation act which opened the market to competition and allowed local phone companies, long distance carriers, cable operators, radio and TV broadcasters, ISPs, and telecom equipment providers to compete in each other's markets. This hastened the race to provide broadband services. This race got further impetus in 1997 by the entry of cable operators using DOCSIS-based cable modems that supported multiple services of data, voice, and video on same co-axial cable inside the customer's premises. We have already discussed HFC-based access in Chapter 1 and an example of HFC access is shown in Fig. 1.12. This competition spurred innovation to provide multiple services on the copper wire pair, and the result is a family of xDSL technologies.

In fact, the concepts that form the foundation for xDSL emerged in the late 1970s. John Cioffi, one of the technology's pioneers, recalls a pivotal meeting[28] at Bell Labs in 1979. There, based on his calculations supported by Shannon's theory,[6] he floated the idea that a four-mile-long twisted pair wire might support transmission rates of 1.5 mbps. His calculations also showed that higher bandwidths could be supported for shorter distances. The variation of bandwidth with distance is shown in Table 2.2.

Table 2.2 Bandwidth Versus Distance for Unshielded Twisted Pair Copper Wire

Bandwidth	Distance
2 mbps	18,000 feet
7 mbps	12,000 feet
12 mbps	10,000 feet
25 mbps	4000 feet
51 mbps	700 feet

At that time, the notion was not readily accepted. There was controversy right from the start due to the fact that higher bandwidth required higher frequencies but as we discussed in Chapter 1, the frequency above 4000 Hz was filtered out in PSTN-based telephone lines. Secondly, there were coils in the line to reduce noise, and these coils had to be removed to support higher frequencies. There was resistance from telephone operators to remove filters and coils because quality of their phone service depended on these two items. There was disagreement about the best transmission method. Fortunately, Cioffi and a handful of others persisted in pursuing what the mathematics implied. As a result, there were mainly two camps. One was led by the telephone operators and the other by Cioffi who had by then joined Stanford University. The industry favored QAM (quadrature amplitude modulation), and Cioffi proposed DMT (discrete multitone). His proposal was based on dividing the 0–1.104 MHz spectrum into 256 small, 4.3125-kHz wide, sub-channels and then using multichannel transmission. This was quite a change compared to what the industry was doing at that point in time. They were using the entire copper wire for one channel on 0–4 kHz frequency. He also had an algorithm to anticipate noise and change channel to avoid or reduce noise. In addition to this approach, Joseph Lechleider of Bellcore suggested asymmetric transmission that included separate downstream bandwidth using more channels and a narrower upstream bandwidth over fewer channels because users downloaded more than uploaded. This would increase speeds sufficiently that movies could be offered over telephone lines. To settle this issue and to decide the winner of QAM and DMT approaches, Bellcore, Bell Atlantic, and Nynex decided to organize an "Olympics." The results of the Olympics shocked the industry. Contrary to expectations, DMT proved to be faster, more efficient, and more flexible than QAM. ANSI was satisfied with the results, and in March, 1993, it chose DMT for its asymmetric DSL (ADSL) standard.[29] The International Telecommunications Union (ITU) and European Telecommunications Standards Institute (ETSI) followed the suit. That was the beginning of ADSL. xDSL got further boost with (1) the emergence of IPTV which as we discussed in Chapter 1 frees up bandwidth and (2) fiber optics which allowed for fiber to the cabinet (FTTC)-based VDSL. In this configuration, copper wire pair distance could be shortened allowing support for higher frequencies resulting in higher bandwidths. Fig. 2.9 shows various combinations of fiber and copper wire pair in the access loop.

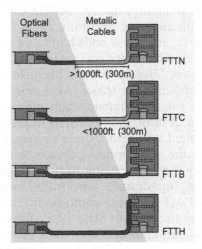

Figure 2.9 Comparison of FTTx access.

Application of FTTC in VDSL technology and that of FTTH in GPON technology is shown in Fig. 2.10. In case of FTTC, optical fiber goes from CO to a DSLAM unit in a street cabinet, and from DSLAM to home there is copper wire pair for a short distance which could support higher frequency without attenuation thus giving higher bandwidth. In case of FTTH or FTTB, there is an optical splitter in the street cabinet which then feeds into fiber going right to home or to basement of a building.

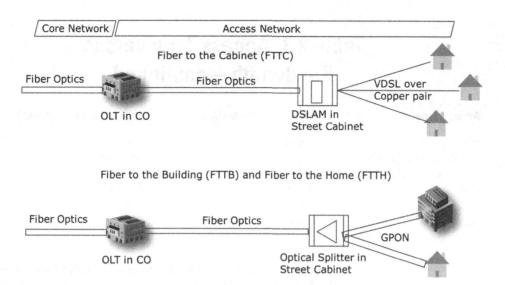

Figure 2.10 Comparison of VDSL and GPON-based access.

As a result of these advances, the widely deployed Copper wire pair from PSTN era is able to provide access to high-bandwidth applications including voice, video, and data services. It is estimated that there are over 500 million xDSL lines worldwide. The history of xDSL shows how few dedicated, knowledgeable folks with tenacity could advance communications technology to enable mass distribution of broadband over the ubiquitous copper wire pair. Table 2.3 gives a summary of upstream and downstream bandwidths for various access types.

Table 2.3 Access Type Versus Bandwidth

Access Type	Upstream Bandwidth	Downstream Bandwidth
Dial-up	56 kbps	56 kbps
HSPA	1400 kbps	1400 kbps
DOCSIS 1.0	2 mbps	2 to 25 mbps
DOCSIS 2.0	27 mbps	38 mbps
DOCSIS 3.0	120 mbps	160 mbps
ADSL	1 mbps	8 mbps
ADSL2	1 mbps	1 mbps
ADSL2+	1 mbps	24 mbps
SHDSL	5.6 mbps	5.6 mbps

Continued

Table 2.3 Access Type Versus Bandwidth—continued

Access Type	Upstream Bandwidth	Downstream Bandwidth
VDSL	15 mbps	55 mbps
VDSL2 long range (12 MHz)	2 to 25 mbps	5 to 50 mbps
VDSL2 short range (30 MHz)	100 mbps	100 mbps
FTTP	500 mbps on PON	500 mbps on PON
WiMax	40 mbps	40 mbps
LTE	10 mbps	20 mbps

Although fiber optics will undoubtedly be the ultimate broadband technology, it is more expensive and time consuming to deploy to every customer's premises. Therefore, fiber optics does not mean the end of xDSL technologies. The transition from the world's legacy access to an all-fiber future will be gradual. This brings us to the development of fiber optics and its use in access loop.

Research on fiber optics traces its origin to an article in 1966 by Kao and Hockham,[30] based on their work at ITTs Standard Telecommunications Laboratory in England, they demonstrated that theoretically a sufficiently pure glass fiber having a low-enough attenuation can be used as a medium for light waves carrying information. Light waves have a much higher frequency and therefore can support higher bandwidths. In 1970, a team led by Robert Maurer at Corning Glass developed the first suitable glass fiber,[31] which Corning then continued to improve. That same year, a team at Bell Labs developed the first room-temperature semiconductor laser, providing a practical pulsing light source suitable for a digital optical system. This was followed in quick succession with test systems in several countries and then with field trials with customers. In 1977, GTE installed a test fiber-optic cable system in Long Beach, California. Old AT&T quickly followed with one in Chicago, and the British Post Office installed a system at Martelsham Heath. While this work was going on, in 1976, J. Jim Hsieh at the Lincoln Laboratory in MIT developed a laser that emitted light at 1.3 μm which was at the same wavelength that a fiber, developed by Masaru Horiguchi at NTT in Japan, could optimally transmit. This improved capacity and reduced loss and made the system more efficient. Other advances followed over the next few years. In 1983, US long distance company MCI, working with Corning, opened a commercial, 1.3-μm, fiber-optic cable system between New York and Washington, which was soon followed with a competitive line by old AT&T. Because fiber-optic transmission was digital, it was well suited for the ever increasing quantity of digital computer data being sent over the world's telephone lines. From the mid-1980s, fiber-optic installations expanded rapidly all over the globe and improved systems followed quickly. Capacity of fiber increased even further with each new innovation and reduced operating costs. For example, the last copper transatlantic cable, TAT-7, which opened in 1978 had a capacity of 4000 calls; the first fiber cable, TAT-8, opened in 1988, increased that by 10 folds. By the late 1990s, new generations of fiber optic systems could carry millions of calls; however, most of what was transmitted was data. In terms of bandwidth, coaxial copper cable carried millions of bits or megabits per second which increased to hundreds of megabits due to fiber-optic cable in early 1980s and that further increased to gigabits in 1990s due to innovations in fiber-optic cables, and in 2000s, fiber optic cables carried terabits of data which has now recently reached petabits level. Fiber optics rendered all previous telephone network transmission media obsolete. The cost of transmitting a phone call to any place on Earth within reach of a fiber-optic cable rapidly approached zero, thus knitting the planet more closely into a single instant communication web, greatly facilitating global commerce. The widespread adoption of fiber optics played a key role in making the global Internet possible. By 2000, copper wire for the most part persisted only in local loops, and microwave systems had been largely decommissioned. And now, the use of fiber optics cable in access loop is also growing though gradually. As shown in Fig. 2.11, there are two possibilities based on fiber-optic cable in local loop. First one is (Fig. 2.11 (a)) with

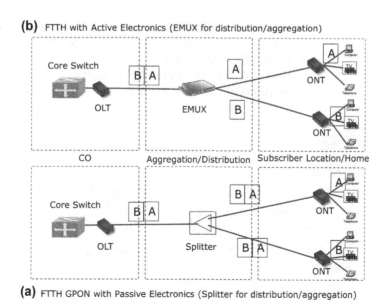

(b) FTTH with Active Electronics (EMUX for distribution/aggregation)

(a) FTTH GPON with Passive Electronics (Splitter for distribution/aggregation)

Figure 2.11 Comparison of point-to-point and GPON-based optical access.

a passive optical device like a splitter that results in GPON and other variations of PON service. Here the same data is sent to all subscribers, and each ONT is required to filter. The second (Fig.11 (b)) possibility has a device with active electronics in the field that switches the data meant for a subscriber. This is expensive compared to the PON approach but provides greater bandwidth and greater control.

Comparison of various PON-based access is shown in Table 2.4. Current trend is toward GPON which uses Ethernet for layer 2 encapsulation. The most common split ratio is 1:32, but if needed, it can be reduced to 1:16 to accommodate higher bandwidth.

Table 2.4 Bandwidth Comparison of Different Optical Access Methods

Type	Specification	Maximum Split Ratio	Downstream Capacity	Upstream Capacity	Layer 2 Encapsulation
APON	FSAN	1:32	622 mbps	155 mbps	ATM
BPON	FSAN/ITU-T G.983	1:32	1.25 Gbps	622 mbps	ATM
EPON	IEEE 802.3ah	1:32	1.25 Gbps	1.25 Gbps	Ethernet
GPON	FSAN/ITU-T G.984	1:64	2.5 Gbps	1.25 Gbps	Ethernet

A typical GPON-based access to a home is shown in much greater detail in Fig. 2.12. This access is shown for delivering a triple play of audio, video, and data services. The backhaul is still shown to have link to PSTN via GR-303 gateway because not all telephone subscribers are on VoIP yet. But that will change in next few years.

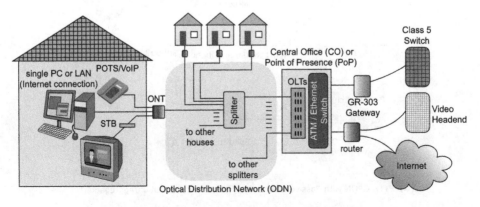

Figure 2.12 Typical GPON Installation in Detail.

The situation shown in Fig. 2.12 is called Greenfield because it is solely based on fiber access for triple play. The situation on the ground is rarely as simple and as clean. Due to installed base, it is always a combination of copper-wire-pair–based POTS service, xDSL-based triple play of voice, video, and data services, or mix of copper-wire-pair–based POTS and fiber optics–based access for data and video. This is called Brownfield. This is shown in Fig. 2.13.

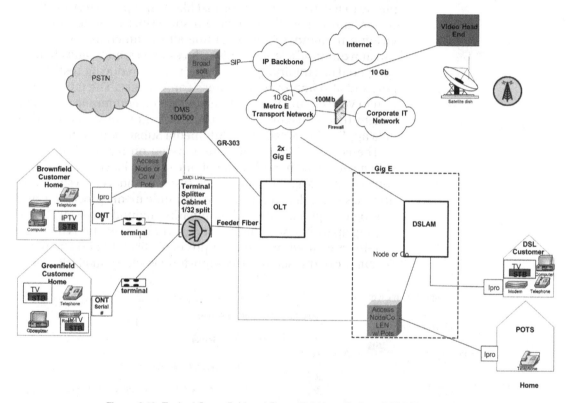

Figure 2.13 Typical Greenfield and Brownfield installation of GPON and xDSL.

Here also, we can see that there are links to PSTN because there are many subscribers who are not yet on VoIP and by implications of Provider of Last Resort (POLR) and Communication Company of Last Resort (COLR), operators have to provide telephone services. However, as we saw in Chapter 1, FCC has agreed to limited trials on scaling down TDM networks. In the next section, we will see an example of a system that is making it possible for voice to be another data application and also providing a framework for other applications.

2.4 Voice as Another Data Application

The key driving force behind transforming voice to become another data application is of course monetary. This saving stems not only from eliminating operating separate voice and data networks but also from combining the current separate IT and support departments for voice and data services. Transforming voice into a data application, commonly known as voice over IP (VoIP) for wireline and voice over LTE (VoLTE) for wireless access, allows phone to receive an IP address just like a laptop gets from a DHCP server, and this enables moving a subscriber's phone from one location to another while retaining all the functions just as one can move around a laptop. This reduces moves, adds, and changes compared to a traditional telephone, thereby reducing cost of support and also improves customer's convenience. The transformation to VoIP or VoLTE also allows for unified communication. The system that is making this transformation of voice into another data application is called IP Multimedia Subsystem (IMS).

The core of IMS architecture is based on understanding the systems involved in traditional phone service like the soft switch, conferencing system, voice mail system, and IVR system and doing a service decomposition to decouple media processing and call-logic requirements and eliminating redundancies and then coming up with best of breed systems where interactions between systems are based on standard protocols like SIP and Diameter. Schematic of this service decomposition is shown in Fig. 2.14.

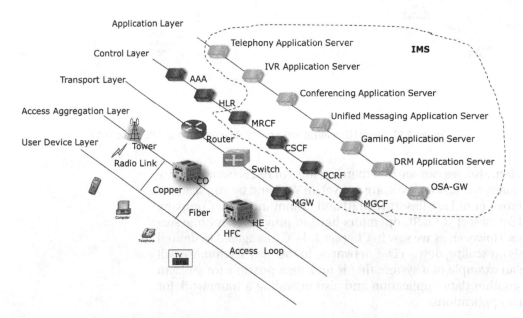

Figure 2.14 Schematic of IP multimedia subsystem.

The schematic shows various layers including user device layer, access aggregation layer, transport layer, control layer, and application layer. Control layer has various servers like the media resource control function (MRCF), call signal control function (CSFC), policy charges and rules function (PCRF), media gateway control function (MGCF), authentication, authorization, and accounting (AAA) and home location register (HLR). In many cases, the functions of AAA and HLR have been combined in one single system called home subscriber server (HSS). The application layer has servers for various telephony and non-telephony applications. It also has an open service access (OSA) gateway for third-party applications servers using Parley-Parley-x APIs.

This service decomposition and use of best of breed systems in IMS provides numerous advantages including (1) reduced capital expenditure, (2) increased port density, (3) reusability of services, (4) lowering integration costs, (5) lowering space requirements, (6) reduced operating costs, (7) increased revenue due to the ability to offer multimedia services, and (8) accelerate introduction of new applications by adding servers in application layer.

An example of VoLTE that uses IMS is shown in Fig. 2.15. Because a subscriber could be roaming, there is a hierarchy of CSCF called S-CSCF for serving CSCF in subscriber's home location, P-CSCF for proxy CSCF in subscriber's current location, and

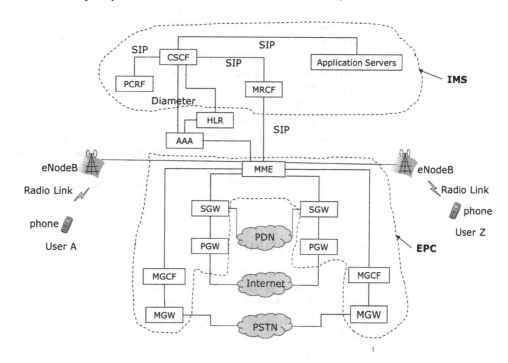

Figure 2.15 Application of IMS in VoLTE.

I-CSCF for interrogating CSCF for the CSCF in the edge location. All user devices have a user agent (UA) to work with IMS, and this UA understands SIP protocol.

When a device is connected to a network supporting IMS, the UA sends SIP signal on control channel. This is received by MRCF via MME. MRCF in turn signals S-CSCF using SIP protocol which in turn determines authenticity of the subscriber by querying HLR, if valid, S-CSCF calls AAA system using Diameter protocol for authentication, authorization, and also to get QoS profile. After this, S-CSCF system calls PCRF to determine policies, rules, and charges and then S-CSCF system sets the QoS and bandwidth requirements on both access and transport networks. With this, the user's device is ready to use IMS services. Let us say user "A" wants to call user "Z" as shown in Fig. 2.15. User "A" is already recognized on the network as described above, so when "A" dials user "Z's" phone number, user "Z" is located by HLR and VLR, and a SIP signal is sent to establish service information using P-CSCF for user "Z". After "Z's" service for IMS is authenticated from his or her HLR/AAA, then a bearer channel is established between "A" and "Z," and "Z's" phone will start ringing and "A" can hear ringing indicating that a connection is established between the two user. When "Z" picks the phone, voice as data flows between the two phones. CSFC then passes the CDR (call detail record) to AAA for billing and accounting purposes. If the user "Z"s phone is not on packet data network (PDN) of the operator then, the call is routed through packet gateway (PGW) over Internet and if the user is on PSTN, then the call is routed by media gateway control function (MGCF) to media gateway (MGW) which then sends the call over PSTN.

Use of IMS is increasing and that is hastening the obsolescence of TDM-based PSTN. As we saw in this description of IMS, it provides for end-to-end QoS which of course depends on the transport network's ability to support that QoS and that is where carrier Ethernet network is playing a critical part and Ethernet in the name indicates that it is the foundation of that network and that is the topic of Chapter 3.

2.5 Chapter Summary

Building on the conclusions from Chapter 1, about emergence of data network as the most dominant network, this chapter described how data network took shape. The factors that influenced this shaping the most are the OSI seven layer model, the ISP-centric architecture blue print from NSF that later morphed into a geography-centric LAN, MAN, RAN, and

WAN based architecture and fiber optics. These were covered in this chapter.

This chapter also described the convergence in backhaul by the emergence of Ethernet as the most popular protocol first at layer 2 and subsequently at layer 1 as well and IP at layer 3 and TCP at layer 4 of the OSI seven layer model.

The diversity in access loop, resulting from installed base of copper wire pair and coaxial cable and newer technologies of fiber optic cable and radio link, has led to rapid growth of customer base. Description of this diversity in access loop was also included in this chapter. Finally, the chapter covered IP multimedia subsystem (IMS) platform which is transforming voice into another data application, thus facilitating move of voice also to data network.

IMS provides for end-to-end QoS which of course depends on the transport network's ability to support that QoS and that is where carrier Ethernet network is playing a critical part and Ethernet in the name indicates that it is the foundation of that network and that is the topic of Chapter 3 where Ethernet landscape will be examined in greater detail.

3

THE ETHERNET LANDSCAPE

I predict the Internet in 1996 will catastrophically collapse.

Remarks by Robert Metcalfe, Co-Developer of Ethernet Protocol, in 1995.

When the Internet did not collapse in 1996, Robert Metcalfe at the sixth International Conference on World Wide Web in 1997 took a printed copy of his prediction and put it in a blender with some liquid to make pulp and drank it. This is one of the rare cases where despite this faux pas, his reputation has grown over the years because of the sound foundation of Ethernet protocol that he and David Boggs developed[11] in 1976. Today over 90% of local area networks (LANs) in the world run on Ethernet because it is fast, easy to install and manage, scales up to bandwidths from 1 Mbps to 100 Gbps and more, and is inexpensive. It has made deep inroads in LAN, broadband access, cloud interconnect, and broadened its penetration in metropolitan area network (MAN), regional area network (RAN), and wide area network (WAN). It is this sound foundation that has enabled Ethernet to still retain its basic frame format even after its spectacular evolution over the years with myriads of associated specifications.

In the first chapter, we covered that data network has emerged as the most dominant network today. In the second chapter, we covered how this data network took shape driven by Open Systems Interconnection (OSI) seven-layer model and the National Science Foundation (NSF) architecture of the Internet which is the network of data networks. In the second chapter, we also covered that Ethernet is becoming the most dominant protocol today in layer 1 and 2 of the OSI seven-layer model. In this chapter, we will cover the fundamentals of Ethernet, its evolution, and its role in Carrier Ethernet networks (CENs).

3.1 Ethernet Protocol for Data Link Layer

Metcalfe and Boggs started work on the development of Ethernet around 1973 and published[11] the Ethernet protocol in 1976 to connect computers in a building or in a campus to form a LAN using coaxial cables. In 1979 they collaborated with a consortium of Digital Equipment Corporation, Intel and Xerox

Peering Carrier Ethernet Networks. http://dx.doi.org/10.1016/B978-0-12-805319-5.00003-4

Corporation known in short as DIX consortium to promote and standardize this Ethernet protocol. The original design is shown in Fig. 3.1A signed by Metcalfe himself. This network was originally called Alto Aloha Network. Its name was changed to Ethernet to make it clear that this system could support any computer and not just Altos computers because just as "ether" carried electromagnetic waves to all radio stations, coaxial cables carried bits to all computers. The DIX specification was submitted to IEEE in 1980, and a second version designated as Ethernet II was submitted in 1982.[32] This implementation is shown in Fig. 3.1B which is widely available on the Internet and is presumed to be drawn by Metcalfe but despite our efforts that could not be established with certainty. The Ethernet II frame format is shown in Fig. 3.1C.

As shown in Fig. 3.1B, in this implementation, the end stations were connected by tapping the shared medium and by using transceivers at the taps. Transceiver was connected by an interface cable to an interface on the end station. There was a controller behind the interface on the end station. The bits coming from end station were arranged in Ethernet frames by the controller. The format of the Ethernet II frame, as shown in Fig. 3.1C, has the address

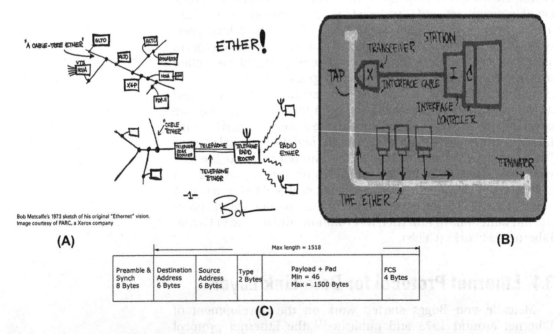

Bob Metcalfe's 1973 sketch of his original "Ethernet" vision.
Image courtesy of PARC, a Xerox company

(A)

(B)

				Max length = 1518		
Preamble & Synch 8 Bytes	Destination Address 6 Bytes	Source Address 6 Bytes	Type 2 Bytes	Payload + Pad Min = 46 Max = 1500 Bytes	FCS 4 Bytes	

(C)

Figure 3.1 Original Ethernet design and Ethernet II frame format. (A) Sketch of original Ethernet design drawn by Metcalfe. (B) Schematic of Ethernet II implementation. (C) Ethernet II frame format.

Image in Fig. 3.1 (A) is courtesy of PARC, a xerox Company http://www.parc.com/content/news/media-library/historical_ethernet_composite_sketch_1973withcc_1.2x8.7_parc.jpg.

of the receiving end station in the 6-byte–long field for destination address (DA). It has the transmitting end station address in the 6-byte–long source address (SA) field. There is a 2-byte–long field called type for identifying type of application using this frame. This is followed by a field for payload which is the actual data to be transmitted. The minimum length of the payload is 46 bytes, and maximum is 1500 bytes. There is a provision for adding some padding if the actual payload is less than the minimum 46 bytes, so that the minimum size of the payload is maintained at 46 bytes. Finally, there is a 4-byte–long frame check sum (FCS) field generated by a special algorithm. These FCS bits are checked by the receiving end station to assure integrity of the received frame. The Ethernet II frame is preceded by a 64-bit–long preamble consisting of 62-bit–long preamble with alternating sequence of 1s and 0s and a 2-bit–long sync character of "11."

The controller shown in Fig. 3.1B was named as the data link layer in the DIX specification, and the interface was called physical layer. DIX specification also divides data link layer into four functional components namely, transmit (Tx) data encapsulation, Tx link management, receive (Rx) link management, and receive data decapsulation. Similarly, the physical layer has four functional components namely, Tx data encoding, Tx channel access, receive channel access, and receive data decoding. For the sake of brevity, the following description only includes data link layer and physical layer. For additional details, please refer to DIX specification.[32]

The process starts when the transmitting end station requests the transmission of a frame and the data link layer of the transmitting end station constructs the frame from the data supplied by the end station and appends a frame check sequence to provide for error detection. The data link layer then attempts to avoid contention with other traffic on the channel by monitoring the carrier sense signal and deferring to passing traffic. When the channel is clear, frame transmission is initiated after a brief delay called interframe gap (IFG) of 9.6 μs to provide recovery time for other data link layers and for the physical channel. The data link layer then provides a serial stream of bits to the physical layer for transmission.

The physical layer performs the task of actually generating the electrical signals on the medium which represent the bits of the frame. The physical layer also monitors the medium and generates the collision detect signal, which, in the contention-free case, remains off for the duration of the frame. It is the physical layer that provides the clock to the data link layer for transmitting bits, and it is the physical layer that converts bits into electrical signal and puts that signal on the interface cable to be transmitted to the transceiver. The transceiver provides the functional electrical and mechanical interface to the shared medium.

The physical layer, before sending the actual bits of the frame, sends the encoded first 62 bits of the preamble to allow the receivers and repeaters along the channel to synchronize their clocks and other circuitry. The preamble is then followed by 2-bit–long sync character to indicate that the Ethernet frame would follow next. It then begins translating the bits of the frame into encoded form and passes them to the transceiver for actual transmission over the medium. When transmission has completed without contention, the data link layer informs the transmitting end station and awaits the next request for frame transmission.

At the receiving end station, the arrival of a frame is first detected by the physical layer, which responds by synchronizing with the incoming preamble and by turning on the carrier sense signal. As the encoded bits arrive from the medium, they are decoded in order to translate the signal back into binary data. The leading bits, up to and including the end of the preamble, are discarded. The receiving end station's physical layer then passes remaining bits to the data link layer.

Meanwhile, the receiving data link layer, having seen carrier sense go on, has been waiting for the incoming bits to be delivered. Receiving data link layer collects bits from the physical layer as long as the carrier sense signal remains on. When the carrier sense signal goes off, the frame is decapsulated for processing.

After decapsulation, the receiving data link layer checks the frame's DA field to decide whether the frame should be received by this station. If so, it passes the contents of the frame to the end station along with an appropriate status code. The status code is generated by inspecting the frame check sequence to detect any damage to the frame enroute and by checking for proper octet-boundary alignment of the end of the frame.

If multiple stations attempt to transmit at the same time, it is possible for their transmitting data link controllers to interfere with each other's transmissions, in spite of their attempts to avoid this by deferring. When two stations' transmissions overlap, the resulting contention is called a collision. A given station can experience a collision during the initial part of its transmission called the "collision window," before its transmitted signal has had time to propagate to all parts of the Ethernet channel. Once the collision window has passed, the end station is said to have acquired the channel and subsequent collisions are avoided, since all other properly functioning end stations can be assumed to have noticed the signal via carrier sense and to be deferring to it. The time to acquire the channel is thus based on the round-trip propagation time of the physical channel.

In the event of a collision, the transmitting station's physical layer first notices the interference on the channel and turns on the collision detect signal. This is noticed in turn by the transmitting data link layer, and collision handling begins. The transmitting data link layer enforces the collision by initiating a backoff algorithm by first transmitting a bit sequence called the jam. This is typically a 32-bit–long frame. This is just the part of a frame that the first end station managed to transmit before the collision occurred. If the collision occurs during preamble, then the jam is returned appended to preamble of 64 bits making a total length of 96 bits which takes 9.6 µs at 10 Mbps, that is, where the IFG of 9.6 µs has come from. If needed, the transmitting data link layer can use a higher IFG with a corresponding decrease in maximum throughput; however, the IFG cannot exceed 10.6 µs. This insures that the duration of the collision is sufficient to be noticed by the other transmitting station(s) involved in the collision. After the jam is sent, the transmitting data link layer terminates the transmission and schedules a retransmission attempt for a randomly selected time in the near future. Since collisions indicate a busy channel, transmitting data link layer attempts to adjust to the channel load by voluntarily delaying its own retransmissions to reduce its load on the channel. This is accomplished by expanding the interval from which the random retransmission time is selected on each retransmission attempt. The retransmission interval is computed using an algorithm called truncated binary exponential backoff algorithm. Here, the station always waits for some multiple "k" of a 51.2-µs time interval, known as a slot time. The station chooses a random number "k" where after first collision, k is chosen from the set (0, 1) and waits for that number of slot time. If there is another collision, it waits again, but this time for a number "k" chosen from the set (0, 1, 2, 3). After three collisions, "k" is chosen from the set (0, 1, 2, 3, 4, 5, 6, 7). After "k" collisions on the same transmission, it chooses its number randomly from $(0\ldots\{2k-1\})$, until $k=10$, when the set is frozen. Eventually, either the transmission succeeds or the attempt is abandoned after 15 unsuccessful attempts, the so-called attempt limit, and the transmitting data link layer gives up and reports a failure to the transmitting end station about the failure because either the channel has failed or has become overloaded.

At the receiving end station, the bits resulting from a collision are received and decoded by the receiving physical layer just as are the bits of a valid frame. The receiving physical layer is not required to assert the collision detect signal during frame reception; however, the assertion of the collision detect signal indicates a true collision in the physical layer. Instead, the fragmentary frames received during collisions are distinguished from valid frames by the receiving data link layer because a collision fragment is always smaller than the shortest valid frame. Such fragments are discarded by the receiving data link layer.

This DIX proposal[32] was approved with minor changes as IEEE 802.3 standard[25] in 1983, and this standard was designated as 10Base5 because it provided 10 Mbps over a thick coaxial cable same as that used in Ethernet II where the signal could be driven up to 500 m. It is important to understand this nomenclature as it will be used for designating variations of Ethernet standards that evolved since 1983. Ten in 10Base5 stands for 10-Mbps bandwidth, "Base" stands for baseband, the fact that the entire medium is used for transmission in contrast to broadband where the medium is divided into several channels at different frequencies for transmission of the signal simultaneously, and 5 represents that the signal could travel without attenuation to a distance of 500 m after that it attenuated. That is why the limit was 500 m. If it was required to extend the distance, then regeneration was needed and up to four repeaters could be used to regenerate and extend the Ethernet resulting in a maximum distance of 2500 m. The standard is designated as IEEE 802.3 because IEEE's subcommittee dealing with LAN began its work on LAN standards in February 1980; therefore, all standards coming out of this subcommittee are numbered starting with 802. Work of this subcommittee was further subdivided into three categories, and all LAN-related specifications started with 802.1, all logical link control (LLC) sublayer–related specifications started with 802.2 and all media access control (MAC) sublayer–related specifications started with 802.3. In 1984 when OSI seven-layer model was published, Ethernet was recognized as one of the protocols for layer 2 and layer 1 for LANs. Since then, it has become the most dominant protocol for these two layers.

Based on the DIX proposal, the IEEE 802.3 standard required that the entire frame length is from a minimum of 64 bytes to a maximum of 1518 bytes. In order to understand how the minimum frame size of 64 bytes was arrived at, let us examine the end-to-end one-way delay time and then arrive at the round-trip end-to-end delay time using the topology of a 10Base5 Ethernet LAN shown in Fig. 3.2.

In order to understand the end-to-end delay time, a simplified topology is shown in Fig. 3.2A. Here, the end-to-end delay time is the sum of serialization time in the interface cable at the source, the processing, queuing, and serialization time at repeater 1, the propagation time in the long coaxial cable plus any point-to-point link, serialization time at repeater 2, and finally serialization time on the interface cable at the destination. The Ethernet II specification recommended slot time to be twice the one-way delay time, in other words, to be equal to the round-trip delay time to

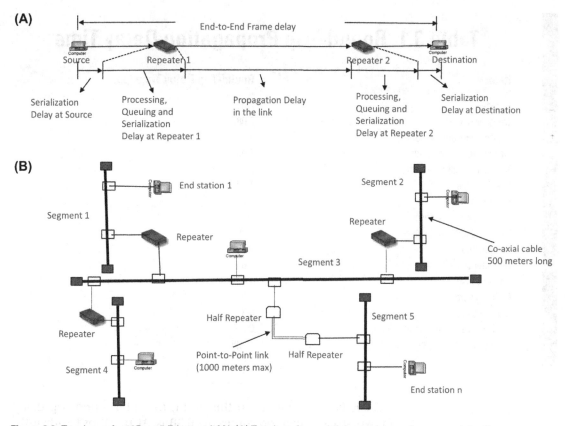

Figure 3.2 Topology of a 10Base5 Ethernet LAN. (A) Topology for explaining end-to-end one way delay time. (B) Topology of 10Base5 for end-to-end round-trip delay time audit.

ensure that source can detect a collision while it is still transmitting. To get this round-trip delay time, the topology[32] of 10Base5 as shown in Fig. 3.2B was used. This topology is based on the interface cable length of 50 and 500 m of the medium in each LAN segment. Repeaters connected the LAN segments, so that signal was not attenuated between sending end station 1 and receiving end station n. The round-trip delay time was obtained by adding time taken, as described previously, to construct a frame, time taken for serialization of the frame and the propagation time of the electrical signal. This audit shown in Table 3.1 is from the DIX specification[32] for Ethernet II and was based on the assumption that the electrical signal travels at speed that is 0.77 times the speed of light which comes to $0.77 \times 300,000,000 = 230,000,000$ m/s. The DIX specification[32] also specifies a minimum propagation speed of 0.65 times the speed of light which comes to about

Table 3.1 Round-Trip Propagation Delay Time

Item	Round-Trip Delay Time (µs)
Encoder	2.0
Transceiver cable	3.08
Transceiver transmit path	2.1
Transceiver receive path	1.95
Transceiver collision path	2.7
Coaxial cable	12.99
Point-to-point link	10.26
Point-to-point link driver	0.40
Point-to-point link receiver	0.40
Repeater path	0.80
Repeater collision path	0.80
Carrier sense	1.0
Collision detect	1.0
Signal rise time	6.3
Collision fragment time tolerance	0.2
Total worst case delay	46.38

195,000,000 m/s. Based on this audit, the total round-trip delay for worst-case scenario comes to 46.38 µs. At 10-Mbps bandwidth, this translates to about 464 bits which was rounded off to 512 bits to be safe. A total of 512 bits are equal to 64 bytes. That is how the minimum size of 64 bytes or 512 bits was chosen.

That minimum frame size at 10 Mbps gives a time slot of 51.2 µs which ensures that the tail of the Ethernet frame has still not left the sending end station when the head reaches the receiving end station which then detects a collision and sends a collision detection signal to the sending end station and is received by the sending end station, so that it can stop transmitting and implement back off algorithm. The maximum frame size of 1518 bytes, on the other hand, was determined based on two factors. First factor was that if the packets are too long, they introduce extra delays to other traffic using the shared medium of Ethernet cable. The second factor was based on a safety device built into the early shared cable transceivers. This safety device was an antibabble system. If the device connected to a transceiver developed a fault and started transmitting continuously, then it would effectively block any other traffic from using that Ethernet cable segment. To protect from this happening, the early transceivers were designed to shut off automatically if the transmission exceeded about 1.25 ms. This equates to a data content of just over 12,500 bits or about 1563 bytes at 10 Mbps. However, as the transceiver used a simple analog timer to shut off the transmission if babbling was detected, therefore a limit of 1518 bytes which includes 1500 bytes of payload and 18 bytes of Ethernet header and FCS, was selected as a safe approximation to the maximum data size that would not trigger the safety device. A detailed analysis of the measured capacity can be found in a report[33] by Boggs et al.

The IEEE 802.3 design and frame format is shown in Fig. 3.3 along with mapping to the original design as well as to the layer 1 and 2 of the OSI seven-layer model.

This standard divided data link layer into two sublayers called LLC layer and MAC layer. LLC allows for many higher level protocols defined by IEEE 802.1 standard to share and use same MAC layer. IEEE 802.2 standard defines LLC layer. LLC uses service access points (SAPs) which represents higher network layer

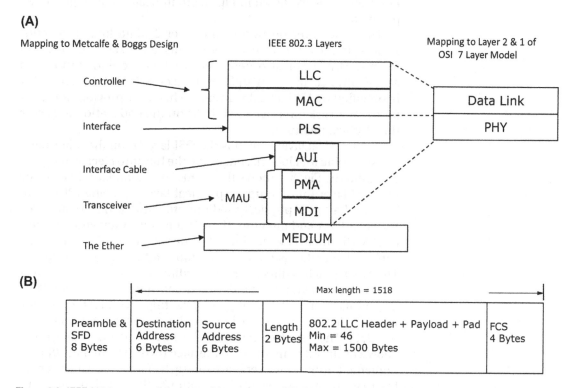

Figure 3.3 IEEE 802.3-approved design and frame format. (A) IEEE 802.3 design of Ethernet interface. (B) IEEE 802.3-approved Ethernet frame format.

protocols being supported by the frames assembled by the MAC layer. There are destination SAP (DSAP) and source SAP (SSAP) and together DSAP and SSAP are called link SAP. Information stored in the SAP fields also determines the type of connection including (1) unacknowledged connectionless, (2) connection oriented, or (3) acknowledged connectionless. In other words, LLC has four main functions:

1. indicates the higher layer protocol using frames at layer 2,
2. provides flow control,
3. provides error recovery, and
4. recovery from loss of connection.

In Ethernet II, SAP fields were referred by Ethertype or Type field (Fig. 3.1). There is another variation of LLC protocol where SAP was replaced by subnet access protocol (SNAP) and is designed to support extended addressing capabilities of transmission control protocol/internet protocol (TCP/IP) and AppleTalk protocol stacks. LLC field is 1- to 2-byte long, and length depends on the higher layer protocol. The IEEE 802.3 standard moved LLC header to payload field and replaced the Type field by a Length field in the Ethernet frame as shown in Fig. 3.3 to indicate total length of the payload.

The MAC layer portion of the layer 2 (data link layer) constructs frames for transmission and analyzes the received frames. MAC layer also determines when the end station can access physical medium and how to access it. MAC functionality is specified by IEEE 802.3 standard. This layer provides a means for the network interface card (NIC) on the end station to access the physical medium.

The physical layer, which is the OSI layer 1, on the other hand is responsible for clocking, encoding the bits into electrical signals and pinouts. Fig. 3.3 shows that IEEE 802.3 standard divided the physical layer into the upper physical layer signaling (PLS) sublayer and a lower physical medium attachment (PMA) sublayer. Between PMA and medium, there is a medium-dependent interface (MDI) sublayer which includes connectors. The PMA and MDI sublayers together is known as medium attachment unit (MAU). The MAU attaches directly to the medium, transmits and receives signals from medium, and identifies collisions. The PLS sublayer is responsible for generating and detecting the Manchester code which ensures that clocking information is transmitted along with the data. The interface between the MAU and PLS sublayers is known as the attachment unit interface (AUI). The AUI in 10Base5 implementation is an interface cable up to 50-m long which carries five twisted pairs connecting the station's NIC (which implements the MAC and PLS) to MAU. In 10Base2 standard which evolved in

1985 for thin coaxial cable and 10Base-T which came out in 1990 for Ethernet over twisted-wire pair, the MAU and AUI are themselves integrated into the NIC connecting directly to the medium. The standard ensured that the MAC sublayer is unchanged in all variations of 10 Mbps 802.3, and its PDUs or frames have a simple structure, shown in Fig. 3.3.

IEEE 802.3 standard replaced 2-bit–long synch with a 1-byte (8-bit)–long start frame delimiter (SFD) while still keeping the overall preamble to 8 bytes. The preamble in this standard consists of 7 bytes of the form 10101010 and is used by the receiver to allow it to establish bit synchronization because there is no clocking information on the medium when nothing is being sent. The SFD is a single byte, 10101011, which is a frame flag, indicating the start of a frame.

The MAC addresses used in 802.3 are always 48-bit (6-byte) long. Each NIC has its own unique address embedded in to the read-only memory chip on the NIC itself. Although the address is hard coded on the NIC, there is an option for the user-defined 2 bytes of this 6-byte–long address field. The embedded address has two components, an IEEE assigned 3 bytes for the NIC card manufacturer's address and 3 bytes assigned by the manufacturer to each NIC card. In the first component, 1 byte is reserved for broadcast or multicast, and the other 2 bytes are for the manufacturer's ID. As a result of these two components, each NIC has a unique address. By normal convention, Ethernet addresses are usually quoted as a sequence of 6 bytes (in hexadecimal) with each byte quoted in normal order but transmitted in reverse order, this arrangement is driven by the transmission order. The mechanism to distinguish universally administered and locally administered addresses is based on setting the second least significant bit of the most significant byte of the address. This bit is also referred to as the U/L bit, short for universal/local bit, which identifies how the address is administered. If the bit is 0, the address is universally administered. If it is 1, the address is locally administered. For example, in address 06-00-00-00-00-00, the most significant byte is 06 (hexadecimal), the binary form of which is 00000110, where the second least significant bit is 1. Therefore, it is a locally administered address. There is also a mechanism to distinguish unicast, multicast, and broadcast frames. If the least significant bit of the most significant octet of a DA is set to 0 (zero), the frame is meant to reach only one receiving NIC, and this type of transmission is called unicast. If the least significant bit of the most significant address octet is set to 1, then it is for multicast transmission. If the most significant address octet of a DA is set to all 1s, then it is a broadcast to all stations on the local network.

The length field is the only one which differs between 802.3 and Ethernet II specifications. In 802.3, it indicates the number of bytes of data in the frame's payload and can be anything from 0 to 1500 bytes. Frames must be at least 64-byte long, not including the preamble, so, if the data field is shorter than 46 bytes, it must be compensated by padding. The reason for specifying a minimum length as explained before lies with the collision detection mechanism. Referring to Fig. 3.3, this includes a minimum payload of 46 bytes plus Ethernet headers and FCS which is 18 bytes thus giving a total of 46 + 18 = 64 bytes = 512 bits.

The last field in the IEEE 802.3 Ethernet frame format is the FCS field which is 4-byte long and is based on a (cyclic redundancy check) CRC-32 polynomial code.

Connecting end stations to a shared coaxial cable was not easy, and also if one end station was removed, it would make the whole LAN inaccessible to other end stations. In 1985, IEEE issued specification 802.3c for 10-Mbps repeaters or hubs to address issues with coaxial cable–based shared medium. And, with the release of 10Base-T standard in 1990, it became possible to implement Ethernet-based LANs on commonly available unshielded twisted-pair (UTP) telephone wires to connect end stations to a hub using RJ45 style connector. This new standard made Ethernet LAN implementation much simpler and faster, and this led to rapid increase of Ethernet adoption in LANs. Fig. 3.4 shows the schematic of both the shared medium and hub-based LANs.

Hub is a multiport device that functions like a shared medium, but it is easier to deploy and improves signal quality

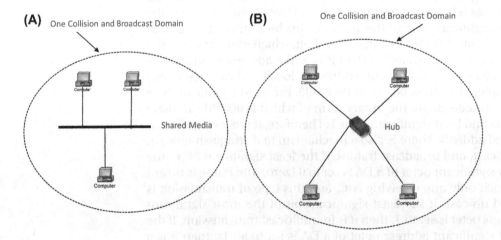

Figure 3.4 Schematic diagram of (A) shared medium and (B) hub-based LAN.

over greater distances. It has no layer 2 functionalities in the sense that it does not look at the layer 2 frames. It is basically a layer 1 device that transmits the bits and just broadcasts them through a shared electrical bus or back plane to all end stations connected to it; therefore, it is a multiport repeater. Although, use of pair of wires allowed separation of transmitted signal from the receiving signal, but the hub's back plane acted as a shared medium, and so collision was still an issue that MAC layer had to deal with by monitoring and implementing backoff algorithm. Fig. 3.4 shows that use of hub did not change the collision and broadcast domains.

As the Ethernet LANs grew in size, the collision issues became critical. This led to breaking up larger LANs into smaller ones. But this led to the desire to interconnect these smaller LANs. This was resolved by the development of bridge. Bridge was built by implementing software on a generic hardware platform. Because of this software-driven intelligence, a bridge is a layer 2 device in the sense that it is able to analyze Ethernet frames and forward the frames to the destination end stations only. Also, because of the RJ45-based connection using UTP wire pair, a bridge supports full-duplex (FDX) mode (ignoring the first 4 years or so of bridges) and not the half-duplex (HDX) mode as the case was with the shared medium. For these reasons, MAC layer on end stations are not required to run carrier sense multiple access/collision detection (CSMA/CD) for collision detection. A bridge is also capable of buffering frames which allows multiple end stations to transmit at the same time. A bridge connects two or more hubs to create a larger LAN. End stations can also be directly connected to a bridge. Each subnetwork of the larger or aggregate LAN is called LAN segment. Every port on a bridge has a MAC address but unlike end station, which only accepts frames addressed to it, a port on the bridge accepts all frames even if it is not addressed to the port. A bridge implements MAC address learning which allows it to gradually build a forwarding database consisting of MAC addresses of all the end stations connected on the LAN and the ports on the bridge by which they can be reached. It builds this table by examining the Ethernet frame whenever it receives it and stores the source MAC address and the port on which the frame arrived. When the bridge receives an Ethernet frame for a destination which is in the forwarding database, then it will send the frame to the port from which the destination can be reached. This type of frames is known as the known unicast frames. If a destination is reachable on the same port as the source, then the bridge discards it because this frame would have been already broadcasted by the hub to which both the source and destination are attached. On the other hand, if the bridge receives an Ethernet frame for a destination that is not in the forwarding database, then it broadcasts the frame to all end stations, except the one from which it received the frame to avoid loop back. This process of flooding unknown unicast frames allows the bridge to determine the port that reaches the destination.

By implementation of LAN segments, MAC learning and intelligently forwarding Ethernet frames, bridge improved the performance of the aggregate LAN by filtering traffic that is local to a LAN segment and forwarding nonlocal traffic to only the correct segment.

The bridge later evolved into Ethernet switch by improving performance further through dedicated hardware instead of software and having large number of ports, so that each end station can be on its own port. The cost also reduced with time, so that Ethernet switch became a cheap device. Now the lines of distinction between a switch and a bridge have gone away, and it is common to use the terminology of switch and bridge interchangeably. Fig. 3.5 shows the LAN arrangement based on a bridge and another LAN arrangement based on an Ethernet switch, and they are functionally the same. As seen from Fig. 3.5, now the collision domains have been

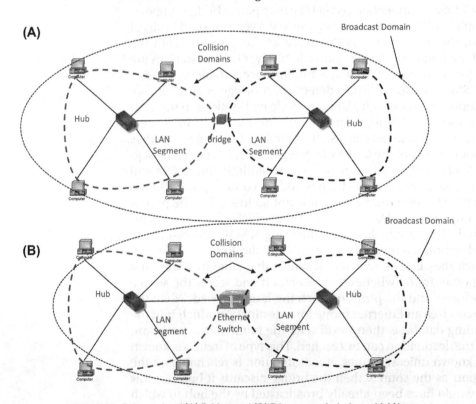

Figure 3.5 Schematic of (A) Bridge and (B) Ethernet switch–based LAN.

broken up and are confined to LAN segments only; however, the broadcast domain still remains intact and is not broken up.

When Ethernet was HDX data could be transmitted in only one direction at a time. With the development of FDX, this situation changed and switches made it easier to exploit FDX mode. In a fully switched network, each node communicates only with the switch, not directly with other nodes. Information travels from node to switch and from switch to node simultaneously. Fully switched networks employ either twisted-pair or fiber optic cabling, both of which use separate medium for transmitting and receiving data. In this type of environment, Ethernet end stations need not implement the collision detection process since they are the only potential devices that can access the medium. In other words, traffic flowing in each direction has a lane to itself. This allows nodes to transmit to the switch as the switch transmits to them in a collision-free environment. Transmitting in both directions can effectively double the apparent speed of the network when two nodes are exchanging information. If, for example, the speed of the network is 10 Mbps, then each node can transmit simultaneously at 10 Mbps.

A switch is also capable of buffering frames. The switch establishes a connection between two segments just long enough to send the current frames. Incoming Ethernet frames are saved to a temporary memory area or buffer in the switch; the MAC address contained in the frame's header is read and then compared to a list of addresses maintained in the switch's forwarding database. For routing traffic, Ethernet switch uses one of three methods:
1. cut-through,
2. store-and-forward, and
3. fragment-free.

Cut-through switch reads the MAC address as soon as a frame is detected by the switch. After storing the 6 bytes that make up the address information, the switch immediately begins sending the frame to the destination node, even as the rest of the frame is coming into the switch. A switch using store-and-forward will save the entire frame to the buffer and check it for CRC errors or other problems before sending. If the frame has an error, it is discarded. Otherwise, the switch looks up the MAC address and sends the frame on to the destination node. Many switches combine the two methods, using cut-through until a certain error level is reached and then changing over to store-and-forward. Very few switches are strictly cut-through since this provides no error correction. A less common method is fragment-free. It works like cut-through except that it stores the first 64 bytes of the frame before sending it on. The reason for this is that most errors, and all collisions, occur during the initial 64 bytes of a frame.

Ethernet LAN switches vary in their physical design. Currently, there are three popular configurations in use: (1) shared memory type of switch stores all incoming frames in a common memory buffer shared by all the switch input and output ports, and the switch then sends frames out via the correct output ports to the destination nodes; (2) a matrix type of switch has an internal grid with the input ports and the output ports crossing each other; when a frame is detected on an input port, the MAC DA is compared to the lookup table in the forwarding database to find the appropriate output port; the switch then makes a connection on the grid where these two ports intersect so that the frames are sent to the destination nodes; and (3) a bus architecture type of switch, where instead of using a grid, a switch uses an internal transmission path consisting of a bus shared by all the ports using time-division multiple access. A switch based on this bus architecture has a dedicated memory buffer for each port and an application-specific integrated circuit (ASIC) to control the internal bus access.

The development of Ethernet switch was a big improvement due to the implementation of learning, filtering, and intelligent forwarding, and it broke up collision domains by confining them to LAN segments; however, it still had the issues due to one large broadcast domain. As the number of end stations increased, bandwidth was consumed by broadcast traffic, multicast traffic, and unknown unicast traffic. Second, any failure in any link would break the aggregate LAN and communications between end stations connected by that failed link would stop. So there was a need for network protection or redundancy in such a way that did not result in loop-back and resultant broadcast storms.

The issue of redundancy was resolved by the Spanning Tree Protocol (STP) which was later enhanced to Rapid STP (RSTP). Working of this protocol is shown in Fig. 3.6. Here, the aggregate LAN is divided into seven LAN segments. Let us assume that in the beginning, switches 2 and 3 are not connected by LAN segment 4. If node 1 wants to transmit frames to node 2, then the only way is to send it through switch 1 via LAN segments 1 and 2.

Now consider a situation where switch 1 has failed. In this case, there is no way for node 1 to transmit to node 2. To avoid this situation, we connect switches 2 and 3 with a LAN segment 4. Now, even if the switch 1 fails, frames can go through LAN segment 4. However this causes an issue. To understand this issue, let us assume that switches 1, 2, and 3 are not aware of node 2. Once the frame comes from node 1, it is added to the forwarding database in each of the switches. Since these switches do not know about node 2, they will broadcast the frame to all the LAN segments that are connected to them. Since switch 3 will get these broadcasts

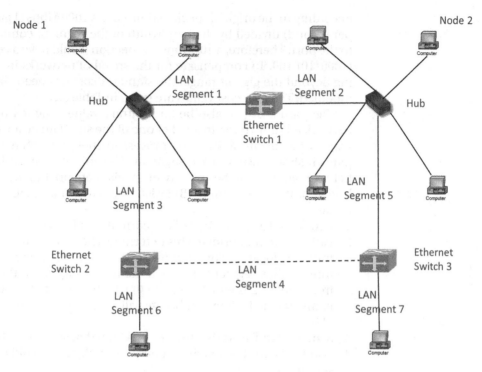

Figure 3.6 Spanning Tree Protocol for redundancy in Ethernet LAN.

from both switches 1 and 2, it will forward broadcast coming from switch 1 to switch 2 and broadcast coming from switch 2 to switch 1. This loop back causes a broadcast storm as the frames are broadcasted, received, and rebroadcasted by each switch, resulting in potentially severe network congestion.

To avoid this broadcast storm while providing redundancy, the STP was developed by Digital Equipment Corporation which has been standardized as the IEEE 802.1D specification. Essentially, a spanning tree uses the spanning tree algorithm (STA), which senses that the switch has more than one way to communicate with a node, determines which way is best, and blocks out the other path(s). Second, it keeps track of the other path(s), just in case the primary path is unavailable. In this protocol, each switch is assigned a group of IDs, one for the switch itself and one for each port on the switch. The switch's identifier, called the bridge ID (BID), is 8-byte long and contains a bridge priority (2 bytes) along with one of the switch's MAC addresses (6 bytes). Each port ID is 16-bit long with two parts: a 6-bit priority setting and a 10-bit port number. Next, a path cost value is given to each port. The cost is typically based on a guideline established as part of IEEE 802.1D.

According to the original specification, cost is 1000 Mbps (1 gigabit per second) divided by the bandwidth of the segment connected to the port. Therefore, a 10-Mbps connection would have a cost of (1000/10) 100. To compensate for the speed of networks increasing beyond the gigabit range, the standard cost has been slightly modified. The new cost values are given in Table 3.2.

The path cost can also be an arbitrary value assigned by the network administrator, instead of one of the standard cost values. Each switch begins a discovery process to choose which network paths it should use for each segment. This information is shared between all the switches by way of special network frames called bridge protocol data units (BPDUs). The parts of a BPDU are as follows:

1. Root BID: This is the BID of the current root bridge.
2. Path cost to root bridge: This determines how far away the root bridge is located. For example, if the data has to travel over three 1 Gbps segments to reach the root bridge, then the cost from Table 3.2 is equal to (4+4+0) which comes to 8. The segment attached to the root bridge will normally have a path cost of 0.
3. Sender BID: This is the BID of the switch that sends the BPDU.
4. Port ID: This is the actual port on the switch that the BPDU was sent from.

All the switches initially send BPDUs to all their neighbor switches, trying to determine the best path between various segments. When a switch receives a BPDU from another switch that is better than the one, it is broadcasting for the same segment, it will stop broadcasting its BPDU out for that segment. Instead, it

Table 3.2 Spanning Tree Cost Values

Bandwidth	Spanning Tree Protocol Cost Value
4 Mbps	250
10 Mbps	100
16 Mbps	62
45 Mbps	39
100 Mbps	19
155 Mbps	14
622 Mbps	6
1 Gbps	4
10 Gbps	2

will store the other switch's BPDU for reference and for broadcast-ing out to inferior segments, such as those that are farther away from the root bridge. A root bridge is chosen based on the results of the BPDU process between the switches. Initially, every switch considers itself the root bridge. When a switch first powers up on the network, it sends out a BPDU with its own BID as the root BID. When the other switches receive the BPDU, they compare the BID to the one they already have stored as the root BID. If the new root BID has a lower value, they replace the saved one. But if the saved root BID is lower, a BPDU is sent to the new switch with this BID as the root BID. When the new switch receives the BPDU, it realizes that it is not the root bridge and replaces the root BID in its table with the one it just received. The result is that the switch that has the lowest BID is elected by other switches as the root bridge. Based on the location of the root bridge, the other switches determine which of their ports have the lowest path cost to the root bridge. These ports are called root ports, and each switch (other than the current root bridge) must have one. It is important to note that a bridge/switch has two or more ports. The one that is connected on the side where the root resides are the root port. A port not facing the root but forwarding traffic at lowest cost from another segment is called designated port. Next, the switches determine designated ports. A designated port is the connection used to send and receive packets on a specific seg-ment. Designated ports are selected based on the lowest path cost to the root bridge for a segment. Since the root bridge will have a path cost of "0," any ports on it that are connected to segments will become designated ports. For the other switches, the path cost is compared for a given segment. If one port is determined to have a lower path cost, it becomes the designated port for that segment. If two or more ports have the same path cost, then the switch with the lowest BID is chosen. Once the designated port for a network segment has been chosen, any other ports that connect to that segment become nondesignated ports. They block network traffic from taking that path, so it can only access that segment through the designated port. By having only one designated port per seg-ment, all looping issues are resolved.

Each switch has a table of BPDUs that it continually updated. The network is now configured as a single spanning tree, with the root bridge as the trunk and all the other switches as branches. Each switch communicates with the root bridge through the root port and with each segment through the designated port, thereby maintaining a loop-free network. In the event that the root bridge begins to fail or have network problems, STP allows other switches to immediately reconfigure the network with another switch acting as the root bridge. This STP process gives the ability to have a complex network that is fault tolerant and yet fairly easy to maintain.

This STP was enhanced in IEEE 802.1w and is called RSTP. In this enhancement, the reconfiguration time due to failure was reduced to 10 s from 50 s that existed for STP. RSTP also supports virtual LANs (VLANs).

The other problem related to one large broadcast domain was resolved by VLANs. Before VLANs, the only way to separate broadcast domains was to use routers. But, routers are layer 3 devices and processing at layer 3 would increase latency. In addition to breaking up broadcast domains, VLANs provided other benefits as well. A VLAN is a logical broadcast domain. Frames sent to a broadcast domain on a specific VLAN are only forwarded to nodes belonging to that VLAN. Initial implementation of VLAN on a switch was proprietary and not based on standards. This hindered adoption of VLANs. Once VLAN implementation was standardized, its adoption grew rapidly.

Standardization of VLANs traces their origin to the IEEE 802.1D standard which describes LAN that include all end stations physically connected to a LAN. In this standard, multicast frames are forwarded to all ports. There was no capability for a switch to determine if the end station connected to a port needed the multicast frames or not. The IEEE 802.1p extension (the lower case "p" indicates this standard is only an extension of 802.1D and not a standalone standard) provided this capability to Ethernet switch to dynamically update the filtering database so that multicast frames are sent to only those ports that have end stations connected to these ports that needed these multicast frames. This extension also provided a capability to prioritize frames to expedite transmission of frames required by time critical applications like voice communications and video conferencing. The specification that standardized VLAN was IEEE 802.1Q–2005 later enhanced by IEEE 802.1Q–2011. It extended the concepts of IEEE 802.1p to provide capabilities to define and support VLANs by defining VLAN tags for identification of VLAN membership and associated priority defined by class of service (CoS). This specification also defined an approach to extend VLANs between switches using trunk lines and multiplexing VLANs over these trunk lines using VLAN tagging. After this specification was issued, the Ethernet frame format was amended in 1998 by IEEE 803.2ac to account for IEEE 802.1Q—defined VLAN tags. Fig. 3.7 shows the VLAN tags as defined by IEEE 802.1Q and the Ethernet frame that includes this VLAN tag. The VLAN tag is 4-byte long. The first 2 bytes are called tag protocol identifier (TPID), and as per 802.1Q–2005, it was set to hexadecimal 8100 represented as 0x8100 or could be set to 0x88a8 (TPID of 0x88a8 was added in the 2011 version of

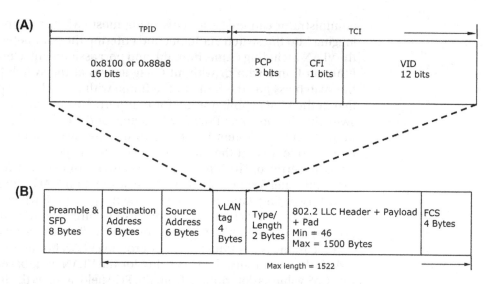

Figure 3.7 VLAN tag and the modified Ethernet frame format. (A) IEEE 802.1Q defined VLAN tag (B) IEEE 802.3ac defined Ethernet frame format to include VLAN tag.

the standard as a result of the 802.1ad amendment. It was not in the 2005 version). Commonly, a VLAN tag with TPID set to 0x8100 is called Customer VLAN tag or C-tag and TPID equal to 0x88a8 is called Service Provider tag or S-tag. It should be noted that an Ethernet frame could have both C-tag and S-tag as we will see in next chapter. The 2-byte–long field following TPID is called Tag Control Information (TCI) field.

This TCI is divided in to three fields. The first field in the Priority Code Point (PCP) field is commonly known as P-bits. It is 3-bit long that gives 2^3 or eight possible values and is used to set the CoS value from 0 to 7, 0 is lowest priority and 7 being highest priority as defined in IEEE 802.1p. Next, there is a 1-bit–long canonical format identifier (CFI) field (this CFI field was eliminated from the 0x8100 tag in the 2011 standard and never existed in 0x88a8 tag. In both cases, it is now designated as Discard Eligibility Indicator (DEI). We will cover it in more detail in Chapters 4 and 5). When it is 0, it indicates an Ethernet frame format, and when it is 1, it indicates a Token Ring frame format. Since majority of the cases, it is Ethernet frame format; it is set to 0. After the CFI field, there is the VLAN-identifier (VID) field which is 12-bit long and gives 2^{12} or 4096 possible VIDs starting from 0 to 4095 identifying which VLAN the Ethernet frame belongs to. This VLAN tag as specified by the IEEE 802.1Q standard is inserted in the Ethernet frame after the SA and before the Length/Type field as shown in Fig. 3.7. To account for this VLAN tag, the Ethernet frame format was modified by the IEEE802.3ac standard. A network administrator can create a VLAN using most switches simply by logging into the switch via Telnet and entering the parameters for the VLAN including name, domain, and port assignment. Once an Ethernet frame with or without C-tag arrives at the switch port, the switch assigns the S-tag to the frame with port VID as specified in the VLAN configuration for that port. This use of S-tag for switching is known as Provider Bridging, and we will cover it in more detail in Chapters 4 and 5. After this, the switch applies the filtering rule, so that the frame is not sent to the port from which it has just arrived. The filtering process also examines the DA field in the Ethernet frame and checks if the destination MAC address is in its forwarding database to determine the port from which destination can be reached. The switch also ensures that the output port is part of the VLAN as defined by the VID. This check ensures that the frame is not transmitted by crossing VLAN boundary. The switch then examines the CoS value for the VLAN and, based on that CoS value as determined from the PCP field, assigns the frame to an output queue or buffer for the output port. The switch also determines if the S-tag is to be assigned or removed when egressing from output port. The S-tag is retained when egressing to a trunk link and is removed when egressing to an access link going to an end station. This process is better explained by an example. Topologies of port-based and extended port-based VLANs are shown in Fig. 3.8. In case of port-based VLAN, four nodes are connected to four ports on the Ethernet switch where nodes 1 and 3 are made members of VLAN 10 and nodes 2 and 4 are members of VLAN 12. So when node 1 sends a broadcast message, it is sent only to members of VLAN 10 that is only to node 3. These port-based VLANs are created by the network administrator manually by logging on to the switch interface and configuring the ports 1 and 3 to be members of VLAN 10 and ports 2 and 4 to be members of VLAN 12. Typically VLANs are identified by numeric assignment as per VID.

The real advantage of VLAN is derived when it is extended to more than one switch as shown in Fig. 3.8B. This is illustrated by having two Ethernet switches connect by a trunk line. Node 1 and 2 are connected to switch 1, and node 3 and 4 are connected to switch 2 and similar to case (A); nodes 1 and 3 are part of VLAN 10 and nodes 2 and 4 are part of VLAN 12. Although frames for both VLANs 10 and 12 travel through the same trunk line between switches 1 and 2, they are kept separate by their respective VLAN tags. Also, explicit tagging of the frames on trunk line with a VLAN identifier VID reduces the processing at the switch because a receiving switch does not have to process the frame to determine VLAN membership. It is important to note that S-tags are relevant

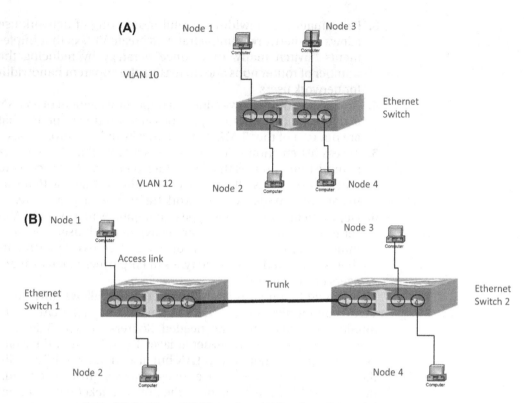

Figure 3.8 Port-based VLAN topology. (A) Port-based VLAN. (B) Extended VLAN.

to ports on the switch only and not to the end stations, they do not know which VLAN they belong to.

We considered the example of port-based VLANs here because they are the most popular VLANs. However, there are other types of VLANs as well including MAC address-based, protocol-based, and policy-based VLANs. As mentioned before, network administrator can create a VLAN using most switches simply by logging into the switch via Telnet and entering the parameters for the VLAN including name, domain, and port assignments. After creation of the VLAN, any network segments or end stations connected to the assigned ports will become part of that VLAN. Some of the common benefits of VLANs are as follows.

1. Security—separating systems that have sensitive data from the rest of the network increases security and decreases unauthorized access.
2. Projects/special applications—allow better management of a project because a VLAN brings all the required nodes together.

3. Performance/bandwidth—careful monitoring of network use allows the network administrator to create VLANs that implements "switch many, route once" strategy by reducing the number of router hops and increasing the apparent bandwidth for network users.
4. Broadcasts/traffic flow—since an important element of a VLAN is the fact that it does not pass broadcast traffic to nodes that are not part of the VLAN, it automatically reduces broadcasts.
5. Access list creation—an access list is a table that the network administrator creates that lists which addresses have access to that network. This allows a network administrator with a way to control who sees what network traffic. It is easy to create.
6. Departments/specific job types—companies may want VLANs setup for departments that are heavy network users (such as multimedia or engineering), or a VLAN across departments that is dedicated to specific types of employees (such as managers or sales people).

It should be noted that an Ethernet switch allows only intra-VLAN communications. If there is a need for inter-VLAN communications, then a router is needed. Routers are layer 3 devices, however, they still use Ethernet at layer 1 and 2 due to the availability of Ethernet not only in LAN but also in MAN. Additionally, the availability of Ethernet over dense wavelength-division multiplexing (DWDM) technology in the optical packet networks has now extended the reach of Ethernet in RAN and WAN as well.

With these fundamental developments involving Ethernet frame format definitions, CSMA/CD process for carrier sensing and collision detection, evolution from shared medium to hub to bridge to switch-based LANs, provision of redundancy based on STP and RSTP, and finally the capability of VLAN creation and handling by Ethernet switches, the adoption of Ethernet for LANs grew rapidly and led to other important enhancements mostly related to increasing bandwidth and distance. This will be covered in the next section.

3.2 Evolution of Ethernet

The 10Base-T standard in 1990 spurred the adoption of Ethernet in LANs because it allowed the use of widely available UTP wire pair with RJ45 connector using hubs, and since use of wire pair provided separate paths for transmitting and receiving signals, collision was not detected due to elevated voltage in the medium, but MAU had to indicate to MAC if it detected simultaneous activities on both Tx and Rx paths because back plane of the hub acted as shared medium, and therefore the MAC layer had to implement backoff algorithm in case of collision. The use of hubs allowed the use of star topology with distance of each LAN link of 100 m and a collision diameter of 2500 m. This standard was based on a physical layer implementation of 10-MHz clock speed and a data path of 1 bit in the PLS layer as shown in Fig. 3.3. It should be noted that the master clock speed for Manchester encoding, which was used for encoding at 10 Mbps, always matches the data speed, and this determines the carrier signal frequency, so for 10-Mbps Ethernet, the carrier frequency is 10 MHz.

With the rapid growth of Ethernet LANs and evolution of Ethernet switch, it became clear that higher bandwidth would be required to support aggregation of 10-Mbps links. The IEEE 802.3u task group was tasked with developing 100 Mbps Ethernet specification. After deliberations, they set the following goals that have served well in the Ethernet evolution:

1. ease of migration and seamless integration with installed base;
2. implementation over widely available UTP wire pair;
3. a 10-fold increase in performance should not increase price by more than a factor of 2;
4. leverage existing technology from (fiber distributed data interface) FDDI; and
5. analyze market research to ensure adoption by various industries and research organizations.

The goal of 100 Mbps bandwidth could be achieved by increasing the clock speed from 10 to 100 MHz but that would reduce the slot time to 5.12 μs from 51.2 μs, and as a consequence, the collision diameter would be reduced from 2500 to 250 m. Both were not acceptable. Therefore, the 100 Mbps standard changed the clock speed to 25 MHz and increased the data path to 4 bits. Increasing clock speed from 10 to 25 MHz was an increase by 2.5 times, and combining this with 4-bit–wide data path gave a $2.5 \times 4 = 10$-fold increase in the bandwidth without reducing the collision diameter substantially. In addition to this, the encoding was changed from Differential Manchester to first non-return-to-zero (NRZ) and later to non-return-to-zero, inverted (NRZI) to meet the needs of increased bandwidth. This encoding/decoding function was moved from PLS layer to a physical coding sublayer (PCS). Also, the MAC layer was decoupled from the physical layer (PHY) by replacing PLS layer with a Medium-Independent Interface (MII) layer which is equivalent to AUI layer as far as functionality is concerned except that MII functions at higher clock speed of 25 MHz and has a 4-bit data path. This MII layer allowed to keep MAC and LLC layer unchanged from 10 Mbps to 100 Mbps evolution. Instead of MII, one could use Reduced MII (RMII) also which uses 50-MHz clock instead of 25-MHz clock and 2-bit data path instead of 4-bit data path. RMII allows the use of eight-pin interface instead of 16-pin interface needed for MII. For discussion purposes, however, we will refer to MII only. To allow for ease of migration, the IEEE 802.3u standard made MII layer support both 10 and 100 Mbps by autonegotiation (AN) sublayer to detect attached end station or device to determine the speed that the device supported and also to detect if device supported duplex or HDX capability. To keep MAC layer unchanged by the introduction of 4-bit data path, the standard introduced a reconciliation sublayer to account for this 4-bit data path. Although with the use of FDX wire pair and switch, CSMA/CD for collision handling became redundant, there still was a need for flow control which thus far was handled indirectly by CSMA/CD function. For this flow control, the standard provided a MAC control sublayer in the MAC layer. This sublayer detected and processed a special MAC control frame or pause frame to affect flow control. Functionally, MAU, which is the combination of PMA and MDI sublayers, remained same and provided an MDI. In addition a physical medium–dependent (PMD) sublayer was added below PMA sublayer so that PMA can get unserialized data from PCS during Tx path and pass serialized data to PMD and receive serialized data on the receive path from PMD and pass unserialized data to PCS. Fig. 3.9 shows the comparison of 10 and 100 Mbps data link and physical layers and also maps to OSI seven-layer model. This figure also shows other standards including 1, 10, 40, and 100 Gbps.

By 1995, 100 Mbps also known as Fast Ethernet was widely adopted and implemented. By that time, the emergence of high-bandwidth applications, which were time critical, required further increase in bandwidth. This time, the IEEE 802.3z task group was tasked to develop the 1000 Mbps or 1 Gbps Ethernet standard. They kept similar goals in mind as the earlier task group had kept while developing 100-Mbps standard. The proposal became standard in 1998 to meet these demands. With the support for VLAN adopted in 1998, the 1-Gbps standard also included support for VLAN and was based on using 8-bit data path and a clock speed of 125 MHz. These two combinations gave a 1000-Mbps bandwidth. However, increasing the clock speed to 125 MHz reduced the slot time to 0.512 µs and collision diameter to 25 m. Both of these were not acceptable. So this time, the standard increased the Tx time to 4096-bit times from 512-bit times by adding a carrier extension while keeping the frame size unchanged. This effectively increased the slot time by a factor of 8 and collision diameter to 200 m. It should be noted that the slot size issue and the frame extension is really only an issue for HDX transmission which is exceedingly rare at 1000 Mbps and somewhat rare today even at 100 Mbps.

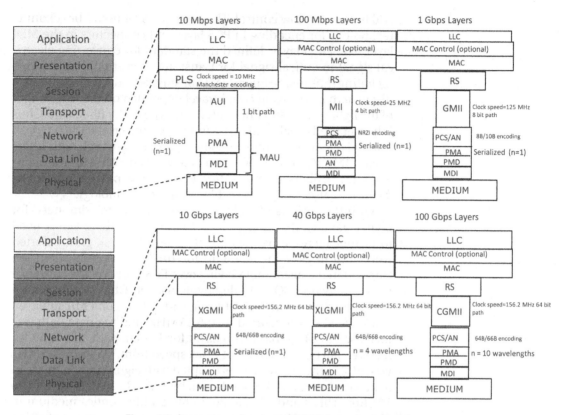

Figure 3.9 Comparison of various IEEE 802.3 standards for Ethernet.

The MII sublayer was modified to gigabit MII (GMII) sublayer to handle higher clock speed and 8-bit data path. This standard also introduced frame bursting to improve performance. Here, if an end station had multiple Ethernet frames waiting for transmission, then the end station would keep the channel occupied by transmitting a control character rather than initiate IFG. This control character was followed by an IFG of 0.096 μs which is the IFG at 1000 Mbps and then transmits the next frame. This control character prevented another end station from capturing the channel. This process is followed to a maximum of 8000 bytes of data before relinquishing the channel. The encoding in PCS sublayer was also changed from NRZI to 8B/10B, and the AN functionality was combined within PCS layer. One Gbps standard also allowed for asymmetric flow control in the sense that switch controlled the flow from an attached device, but the device was not allowed to control traffic from switch. This mechanism allowed buffering on the device in case of flow control and switch did not have to add buffers for flow control. The physical layer used fiber channel technology for encoding in PCS layer and connectors in the MDI layer. The PMD layer defined the standard for converting electrical signal to optical signal for transmitting over fiber optic cable including multimode fiber (MMF) or single-mode fiber (SMF). The distance that the signal could travel depended on the wavelength of the light source, fiber diameter, and type of the fiber. One-Gbps Ethernet used Gigabit Ethernet Interface Converter (GBIC) type of transceivers which supported short- and long-wavelength lasers using MMF (62.5- and 50-µm diameter) and SMF (10-µm diameter) optical fibers as well as short copper wire–pair physical interfaces. This also was based on fiber channel technology.

With the success of 1 Gbps standard and increasing need for higher bandwidths to aggregate 1-Gbps links, there was a need to further increase bandwidth. IEEE 802.3ae task group started work, and in 2002 the standard for 10 Gbps was published. Ten Gbps offered considerable saving over Ethernet over synchronous optical network (SONET) deployments. Ten Gbps was based on 64B/66B encoding in PCS sublayer, and in 10-gigabit MII sublayer, clock speed was increased to 156.25 MHz, and the data path was increased to 64-bit path with 32 bit for Tx data path and 32 bit for receive data path. Increase of clock speed from 125 to 156.25 MHz was a 1.25 times increase this coupled with eight times increase in data path increased the bandwidth by $1.25 \times 8 = 10$ times from 1 to 10 Gbps. This standard operated only on fiber optics medium in FDX mode with 1310- and 1550-nm lasers giving a range of 40 km which extended the reach of Ethernet from LAN to MAN. For larger distances covered by WAN, this standard also provided for a WAN Interface Sublayer (WIS) that formatted the frames to be SONET compatible, so that SONET infrastructure could be used for transport over WAN. This standard also needed a new PMD layer because fiber channel technology–based PMD sublayer used in 1-Gbps standard could not be used. Later, this 10-Gbps standard was modified to support copper wire pair for up to 100 m for data center applications and also to support transport over DWDM with Reconfigurable Optical Add–Drop Multiplexer (ROADM) so that frames could be sent over RAN and WAN without using SONET technology and that provided considerable cost savings. The GBIC transceivers and (10 gigabit small form factor pluggable (X is Roman character for 10)) XFP transceivers were replaced with small-form factor pluggable (SFP) and SFP+ type of transceivers.

With the spread of 10-Gbps Ethernet, stage was set for another 10-fold increase in bandwidth to aggregate 10-Gbps links and also to provide support to increasing data center applications due to the onset of cloud-based applications. The IEEE 802.3ba task group started work on this 40 and 100 Gbps standard in 2008 and published the standard in 2010. The deployment of these 40 and 100 Gbps switches has commenced just now in 2015. The layers for these are also shown in Fig. 3.9. The increase in bandwidth was achieved by retaining 64B/66B encoding, and 156.25-MHz clock speeds in 40- and 100-gigabit MII sublayers from 10 Gbps standard but by increasing the number of lanes or wavelengths in PMA layer to 4- for 40-Gbps and 10- for 100-Gbps bandwidths. This standard also provides for the use of SFP, SFP+, and QSFP (Quad SFP, i.e., 4 SFP transceivers in one) type of transceivers. This standard removed the WIS layer (the layer that provided SONET support) so that Ethernet could use DWDM for WAN applications.

By tweaking clock speeds, data bit path, encoding/decoding methods, slot time, number of lanes or wavelengths for serialization/deserialization, and by leveraging improved ASIC chip based hardware, and enhancements to software, the IEEE 802.3 task groups were able to increase bandwidth from 10 Mbps to 100 Gbps while retaining the original Ethernet frame format shown in Fig. 3.7. Table 3.3 below gives a list[25] of Ethernet evolution, and it is very interesting to track all the important developments related to Ethernet that is so profoundly changing the data networks. It also shows that the work has started on 400 Gbps. Table 3.3 also shows that the standard using multiple 25/50Gbps lanes is expected by 2017. There are already talks on starting the work on developing 1-Tbps (Terabit per second) Ethernet standard. So far, the basic Ethernet frame format shown in Fig. 3.7 has remained unchanged, and it will be interesting to see if it will remain so in the case of 1 Tbps as well.

Table 3.3 Evolution of Ethernet

Serial Number	Year	Name	IEEE Standard	Description
1	1973	Experimental Ethernet	–	2.94 Mbit/s over coaxial cable (coax) bus
2	1982	Ethernet II (DIX v2.0)	–	10 Mbit/s over **thick coax**. Frames have a type field. This frame format is used on all forms of Ethernet by protocols in the Internet protocol suite.

Continued

Table 3.3 Evolution of Ethernet—continued

Serial Number	Year	Name	IEEE Standard	Description
3	1983	10BASE5	802.3	10 Mbit/s over thick coax with a maximum distance of 500 m. Same as Ethernet II (above) except type field is replaced by length, and an 802.2 LLC header follows the 802.3 header. Based on the CSMA/CD process.
4	1985	10BASE2	802.3a	10 Mbit/s over thin coax (a.k.a. thinnet or cheapernet) with a maximum distance of 185 m
5	1985	10BROAD36	802.3b	10-Mbps baseband Ethernet over three channels of a cable television system with a maximum cable length of 3600 m
6	1985	Repeater	802.3c	10-Mbit/s (1.25-MB/s) repeater specs
7	1987	FOIRL	802.3d	Fiber-optic inter-repeater link
8	1987	1BASE5	802.3e	Star LAN (use of hubs)
9	1990	10BASE-T	802.3i	10 Mbit/s over twisted pair
10	1993	10BASE-F	802.3j	10 Mbit/s (1.25 MB/s) over fiber-optic
11	1995	100BASE-TX, 100BASE-T4, 100BASE-FX	802.3u	Fast Ethernet at 100 Mbit/s w/ autonegotiation. TX is two or four pair category 3 or higher unshielded twisted-pair cable. FX is over two multimode optical fibers. T4 is over four pairs of category 3 or higher unshielded twisted-pair cable.
12	1997	–	802.3x	Full duplex and flow control; also incorporates DIX framing, so there is no longer a DIX/802.3 split.
13	1998	100BASE-T2	802.3y	100-Mbps baseband Ethernet over two pairs of category 3 or higher unshielded twisted-pair cable
14	1998	1000BASE-X	802.3z	1-Gbit/s Ethernet over fiber-optic. X is a generic name for 1000-Mbps Ethernet systems

Table 3.3 Evolution of Ethernet—continued

Serial Number	Year	Name	IEEE Standard	Description
15	1998	–	802.3-1998	A revision of base standard incorporating the above amendments and errata
16	1999	1000BASE-T	802.3ab	**1-Gbit/s (1000 Mbps) baseband Ethernet over four pairs of category 5 unshielded twisted-pair cable**
17	1998	Includes 802.1Q VLAN Tags	802.3ac	Max frame size extended to 1522 bytes (to allow "Q-tag") the Q-tag includes 802.1Q **VLAN information** and 802.1p priority information.
18	2000	Includes LAG	802.3ad	**Link aggregation** for parallel links, since moved to IEEE 802.1AX
19	2002	–	802.3-2002	A Revision of base standard incorporating the three prior amendments and errata
20	2002	10GBASE-SR, 10GBASE-LR, 10GBASE-ER, 10GBASE-SW, 10GBASE-LW, 10GBASE-EW	802.3ae	**10-gigabit Ethernet over fiber**
21	2003	–	802.3af	Power over Ethernet (15.4 W)
22	2004	–	802.3ah	**Ethernet in the first mile**
23	2004	10GBASE-CX4	802.3ak	10 Gbit/s (1250 MB/s) Ethernet over twin-axial cables.
24	2005		802.3-2005	A Revision of base standard incorporating the four prior amendments and errata.
25	2006	10GBASE-T	802.3an	10 Gbit/s Ethernet over unshielded twisted pair
26	2007	Backplane Ethernet	802.3ap	Backplane Ethernet (1 and 10 Gbit/s over printed circuit boards)
27	2006	10GBASE-LRM	802.3aq	10 Gbit/s Ethernet **over multimode fiber (MMF)**
28	2006	–	802.3as	Frame expansion
29	2009	–	802.3at	Power over Ethernet enhancements (25.5 W)

Continued

Table 3.3 Evolution of Ethernet—continued

Serial Number	Year	Name	IEEE Standard	Description
30	2006	–	802.3au	Isolation requirements for power over Ethernet (802.3-2005/Cor 1)
31	2009	EPON	802.3av	10 Gbit/s EPON
32	2007	–	802.3aw	Fixed an equation in the publication of 10GBASE-T (released as 802.3–2005/Cor 2)
33	2008	–	802.3-2008	A revision of base standard incorporating the 802.3an/ap/aq/as amendments, two corrigenda and errata. Link aggregation was moved to 802.1AX.
34	2010	–	802.3az	Energy-efficient Ethernet
35	2010	40 & 100 Gbps	802.3ba	**40 and 100 Gbit/s Ethernet.** Forty Gbit/s over 1-m backplane, 10-m Cu cable assembly (4×25 Gbit or 10×10 Gbit lanes) and 100 m of MMF, and 100 Gbit/s up to 10 m of Cu cable assembly, 100 m of MMF, or 40 km of SMF, respectively
36	2009	–	802.3-2008/Cor 1	Increase pause reaction delay timings which are insufficient for 10 Gbit/s (workgroup name was 802.3bb)
37	2009	–	802.3bc	Move and update Ethernet-related TLVs (type, length, values), previously specified in Annex F of IEEE 802.1AB (LLDP) to 802.3.
38	2010	–	802.3bd	Priority-based flow control. An amendment by the IEEE 802.1 data center bridging task group (802.1Qbb) to develop an amendment to IEEE standard 802.3 to add a MAC control frame to support IEEE 802.1Qbb priority-based flow control.
39	2011	Ethernet MIB consolidation	802.3.1	MIB definitions for Ethernet. It consolidates the Ethernet-related MIBs present in Annex 30A&B, various IETF RFCs, and 802.1AB annex F into one master document with a machine readable extract (workgroup name was P802.3be).

Table 3.3 Evolution of Ethernet—continued

Serial Number	Year	Name	IEEE Standard	Description
40	2011	–	802.3bf	Provide an accurate indication of the transmission and reception initiation times of certain packets as required to support IEEE P802.1AS.
41	2011	–	802.3bg	Provide a 40 Gbit/s PMD which is optically compatible with existing carrier SMF 40 Gbit/s client interfaces (OTU3/STM-256/OC-768/40G POS).
42	2012	–	802.3-2012	A revision of base standard incorporating the 802.3at/av/az/ba/bc/bd/bf/bg amendments, a corrigenda, and errata.
43	2014	–	802.3bj	Define a 4-lane 100 Gbit/s backplane PHY for operation over links consistent with copper traces on "improved FR-4" (as defined by IEEE P802.3ap or better materials to be defined by the task force) with lengths up to at least 1 m and a 4-lane 100-Gbit/s PHY for operation over links consistent with copper twin-axial cables with lengths up to at least 5 m.
44	2013	–	802.3bk	This amendment to IEEE Standard 802.3 defines the physical layer specifications and management parameters for EPON operation on point-to-multipoint passive optical networks supporting extended power budget classes of PX30, PX40, PRX40, and PR40 PMDs.
45	2015		802.3bm	100G/40G Ethernet for optical fiber
46	2014	1000BASE-T1	802.3bp	Gigabit Ethernet over a single twisted pair, automotive and industrial environments
47	2016	40GBASE-T	802.3bq	For 4-pair balanced twisted-pair cabling with 2 connectors over 30 m distances
48	2017	400 Gbps over optical fiber	802.3bs	**400 Gbit/s Ethernet** over optical fiber using multiple 25G/50G lanes

Continued

Table 3.3 Evolution of Ethernet—continued

Serial Number	Year	Name	IEEE Standard	Description
49	2017		802.3bt	Power over Ethernet enhancements up to 100 W using all 4-pairs balanced twisted-pair cabling, lower standby power and specific enhancements to support IoT applications (e.g. lighting, sensors, building automation).
50	–	100BASE-T1	802.3bw	100 Mbit/s Ethernet over a single twisted pair for automotive applications
51	2015	–	802.3-2015	802.3bx—a new consolidated revision of the 802.3 standard including amendments 802.3bk/bj/bm
52	2015		802.3by	25G Ethernet
53	2017		802.3bz	2.5 gigabit and 5 gigabit Ethernet over Cat-5/Cat-6 twisted pair—2.5GBASE-T and 5GBASE-T
54	TBD	1 Tbps and beyond	TBD	1 Tbps and beyond

EPON, Ethernet passive optical network; *IoT*, internet of things; *LLDP*, link layer discovery protocol; *MIB*, management information base; *RFC*, request for comments; *TBD*, to be determined.

Now that we have explored the definition of the Ethernet and its evolution, it will be helpful to briefly examine the components related to hardware of a typical Ethernet switch because it is the hardware that forms the data network and it is the network that moves bits from source to destination and on this movement of bits, ride the services needed by customers.

3.3 Components of an Ethernet Switch

In this section, we will briefly examine the hardware-related components of a typical Ethernet switch. There is plenty of information available on the Web and in the manufacturers' product catalogs, and therefore the description here will be limited to high level in order to describe functional components including data plane, control plane, switch fabric, and back plane. The description will also cover layout of a switch processor and functional components of an ASIC chip. This section also includes some examples of Ethernet switch at device level and information about popular transceivers and connectors.

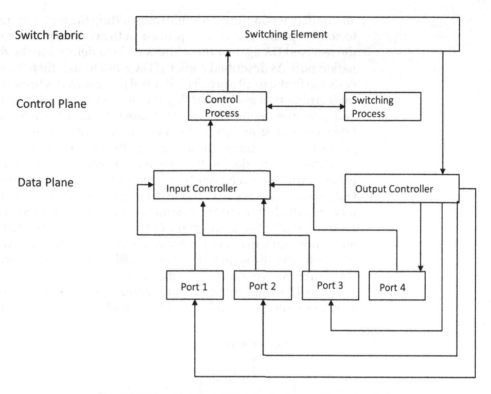

Figure 3.10 Functional components of a typical Ethernet switch.

Fig. 3.10 shows functional levels of an Ethernet switch consisting of data plane, control plane, and switch fabric. The data plane
has input and output controllers. The control plane consists of
the control and switch processes to analyze the frames to determine how to handle the incoming Ethernet frames and using the
switching fabric for connecting input and output ports so that the
bits are moved from source to the destination. The control plane
also includes implementation of the address learning, address
searching, forwarding table and management, and administrative
functions to minimize manual interventions during processing.
The switch fabric, as the name suggests, interconnects ports so
that bits could move from input port to output port. In addition,
there is a back plane to which the data plane, control plane, and
switch fabric connect. In other words, the back plane provides a
bus for the bits to move from one plane to another.

Input controller in the data plane includes PHY and MAC layer.
Its functions are to receive data frames and to filter out invalid
frames (frames that are shorter than 64 bytes or with CRC error).
It transitions between cut-through and store-and-forward modes

and buffers incoming data while transmitting the received frames to control plane. The control process in the control plane verifies the received DA against the address table to determine the destination port. As described earlier, if DA is not found, then it broadcasts the frame to all ports. This control process also is responsible for learning process by entering the new SA in the address table and performs aging process to remove outdated SA from the table. Once destination port is determined, this control process performs the treatment for unicast, multicast, and broadcast and then forwards the data to the switch fabric which is the switching element which then forwards the frames to the output controller. Output controller receives frames from the switching element and forwards the frames to the destination port based on the header information. It also monitors the output resources for flow control and sends signal to the switching element if congestion is detected in which case the switching element will send a PAUSE frame to the source port to suspend the data transfer.

Fig. 3.11 shows more granular details of the functions of an Ethernet switch at chip level. This example includes support for

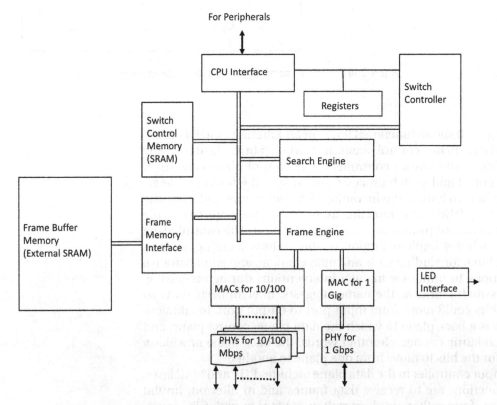

Figure 3.11 Implementation of functions of an Ethernet switch at the chip level.

N number of 10/100 autosensing fast Ethernet ports with MII sub-layer in the physical layer shown as PHY in Fig. 3.11. It also shows a single gigabit Ethernet port with GMII sublayer in the PHY operating at full wire speed and FDX layer 2 switching. Internal switch database for the configuration shown in Fig. 3.11 maintains up to 2000 MAC addresses; however, with external buffer memory, it supports up to 32,000 MAC addresses today on most switches. This implementation includes IEEE 802.3x standard–based flow control and support of 256 ports and also support for IEEE 802.1 Q-based ID tagged VLANs including VLAN tag insertion and extraction.

When Ethernet frame data is received from a port, it is temporarily stored in the MAC receive (Rx) FIFO (first in, first out) until the frame engine moves it to the chip's external memory one granule (128-byte-or-less fragment of frame data) at a time. The frame engine then issues the search engine a switching request that includes the source MAC address, the destination MAC address, and the VLAN tag. After the search engine resolves the address using forwarding database, it transfers the information back to the frame engine via a switching response that includes the destination port and frame type (e.g., unicast or multicast). Switch controller is designed to implement highly efficient management functions for the switching hardware, minimizing the management activity intervention during frame processing. It also has implementation for two modes of operation, namely, cut-through mode and store-and-forward mode. It also allows for individual port configuration as FDX or HDX.

Assuming that there are 8×100 Mbps ports and 1×1000 Mbps (1 Gbps) port in Fig. 3.11 and all ports are FDX, the aggregate throughput supported is equal to $8 \times 2 \times 100 + 1 \times 2 \times 1000 = 3.6$ G bps in equivalent HDX mode throughput. The multiplier 2 is for separate Tx and Rx links in duplex mode. Filtering and forwarding decision must be made in a short time for cut-through mode of operation. Forwarding decision time is obtained by dividing the interarrival time for minimum frames of 64 bytes each at full load by the number of ports. Interarrival time between frames is the time interval between start of two frames in a back-to-back Tx mode from a given port. The shortest interarrival time for minimum Ethernet frame size of 64 bytes is the sum of slot time, IFG, and time for sending the 7-byte preamble and 1-byte SFD. In other words, it is the time taken for transmitting one minimum frame of 64 bytes plus IFG time to transmit 12 bytes (96 bits) and time to transmit 7-byte–long preamble plus 1-byte–long SFD at 100 Mbps or at a rate of 0.01 μs per bit. This comes to $(64 + 12 + 8) \times 0.01 \times 8 = 6.72$ μs of interarrival time between back-to-back 64-byte frames.

The sum of 8×100 Mbps ports and 1 Gbps port is equivalent to $8 + 10 = 18$ ports at 100 Mbps. Therefore, the chip has to arrive at the forwarding decision in $(6.72/18)\,\mu s = 0.37\,\mu s$ or 370 ns.

There is plenty of time for an ASIC chip to accomplish this because, for example, a 733 MHz with 90-nm gate, 64-bit–wide ASIC chip processing three instructions per cycle will require $1/(733 \times 106 \times 3) = 0.45$ ns per instruction and assuming that it takes four instructions to process one forwarding decision including loading the DA into a register in the CPU chip, loading the SA into a register in the CPU chip, comparing DA with forwarding table in the CPU chip, and finally sending the input and output ports to switch fabric. Therefore the chip will require $0.45 \times 4 = 1.8$ ns. But we have 370-ns time from network perspective, so the ASIC chip has plenty of time to process. This time will reduce even more for ASIC chips with GHz clock speeds. For example, a 10-Gpbs Ethernet switch with 132 ports will require an interarrival time obtained by adding slot time, IFG time for 96 bits, and time taken to transmit 7 bytes of preamble and 1 byte of SFD at 10 Gbps. Also, remembering from Section 3.2 that the slot time has been increased to 512 bytes for 10 Gbps switch from 512 bits, we get $(4096 + 96 + 64)/(10 \times 10^9) = 425$ ns. Since this switch has 132 ports, therefore the chip has to arrive at the forwarding decision time in $425/132 = 3.22$ ns from network perspective. Now, assuming that this Ethernet switch uses a 2.6-GHz ASIC chip and assuming three instructions per cycle and as before we need four instructions for forwarding decision, the chip takes only 0.513 ns from ASIC-chip hardware perspective. Therefore, there is plenty of time to accomplish this. These calculations are based on typical single processor switch architecture; however, modern switch technology comprises multiple parallel processing blocks making switch processing so fast that the switch is really a pipeline.

Details of a typical ASIC processor's functions are shown in Fig. 3.12. As we discussed in the last section, each port on the switch is connected to the medium by its PHY which is the layer 1 of the OSI seven-layer model, and at functional level, it is part of the data plane. This layer then sends the data in the form of frames to MAC layer which is the layer 2 of the OSI model, and it is also part of the data plane. Therefore, at implementation level, single chip devices manage both the PHY functions including clocking, encoding, and serialization and MAC layer functions including receiving and transmitting frames, switching between cut-through mode and store-and-forward mode, filtering invalid frames, buffering incoming frames, and transmitting buffered frames to control process. All these functions are implemented on a single chip. The switch control functions including frame engine, search engine, learning, forwarding, and switch fabric are implemented by an ASIC switch or by a Field-Programmable Gate Array (FPGA) chip. Fig. 3.12 also shows a port for the administrator to login for configuration related activities and also for connecting peripherals to the Ethernet switch.

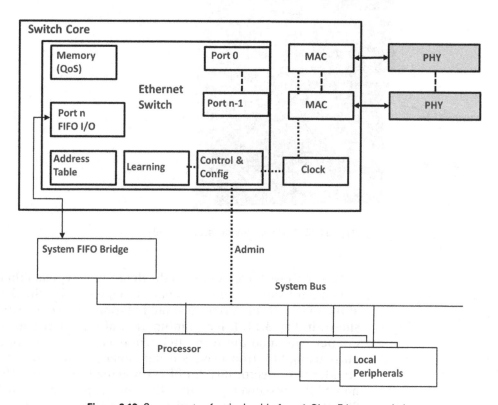

Figure 3.12 Components of a single chip for a 1-Gbps Ethernet switch.

Fig. 3.13 shows, just as an example, partial view of a typical two-port Ethernet switch board with integrated PHY and MAC chip, clock, (synchronous dynamic random access memory) SDRAM, SRAM (static random access memory), FLASH, ASIC, or FPGA switch core and port for peripherals. The integrated single-chip device manages both PHY and MAC layer functions and provides an interface to the host processors by using on-chip command and status registers and a shared host-memory area. They are optimized for LAN-on-motherboard designs, enterprise networking, and Internet appliances that use the Peripheral Component Interconnect (PCI) or PCI-X bus.

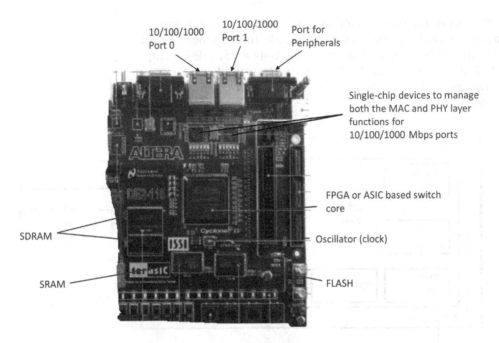

Figure 3.13 A typical Ethernet switch board.

Now that we have examined the details of a typical Ethernet switch hardware at chip and board level, let us see the details at the device level. A typical small five-port Ethernet switch is shown in Fig. 3.14. It has combination of three ports for UTP Copper wire medium using RJ45 connectors and two optical ports using SFP transceivers using Lucent Connector (LC) or Standard Connector (SC) for fiber optic medium. It also shows how SFP transceiver goes into the optical port and how this SFP transceiver is connected to optical fiber medium using LC or SC connectors. Fig. 3.14 also shows the light source on the Tx slot and photodetector on the receive slot of the SFP transceiver.

Fig. 3.15 shows another example of an Ethernet switch. This is a much higher capacity Ethernet switch primarily used on the aggregating and peering networks and also in large data centers that host cloud applications on hundreds or thousands of servers.

Fig. 3.15 shows the craft interface on top with (light-emitting diode) LED light, and the upper and lower fan tray for cooling air. There are several slots to accommodate various types of ports both for copper and fiber optic cables. There are slots for switch control board and switch fabric. Not shown in Fig. 3.15 is a back

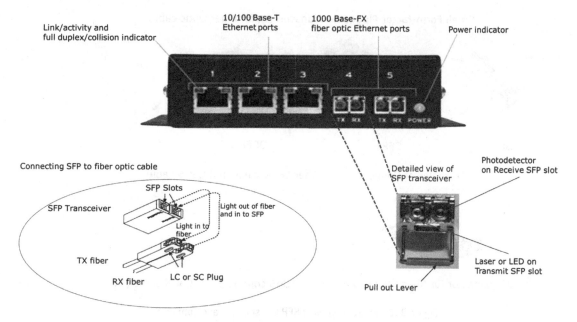

Figure 3.14 Ethernet switch with port-level details.

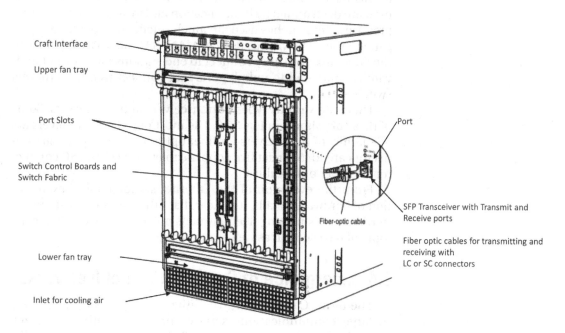

Figure 3.15 Box-level details of a typical Ethernet switch.

Small Form-factor Pluggable Transceivers for Fiber Optic cable

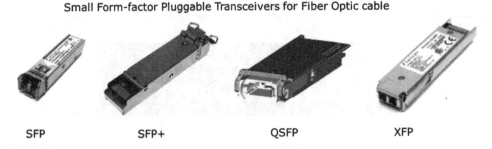

SFP   SFP+   QSFP   XFP

Examples of Connectors for Fiber Optic cable and Cat 5 cable

LC connector for fiber optic cable   RJ45 connector for Cat 5 cable (Copper)

Figure 3.16 Different types of SFP transceivers and connectors.

plane which is inside the box. All other cards in various slots attach to this back plane. The type and shape of Ethernet switch to be used are determined by its application and location. For customer premises, the switches used are much smaller compared to aggregating Ethernet switches used in MAN, RAN, and WAN. Similarly, the downlink ports that connect to end stations have lower bandwidths compared to uplink ports that connect to other Ethernet switches.

There are various types of transceivers and connectors. Some of these are shown in Fig. 3.16. The most popular transceivers are SFP family of transceivers for fiber optics and most popular connectors are RJ45 for cat five copper cables and LC and SC connectors for fiber optic cables.

From the evolution of the Ethernet standards and examination of hardware details, it is clear that Ethernet is not just limited to LANs but is making deep in-roads outside of LANs. That is the topic of next section as well as next chapter.

3.4 Emergence of Carrier Ethernet Networks

The term "Carrier" in carrier Ethernet networks (CEN) refers to large communications services providers with wide reach through their global networks. These carriers provide audio, video, and data services to residential as well as to business or enterprise customers. Enterprise customers generally have the need for other services as well which are very specialized including network transport services for businesses with locations that are geographically dispersed over the globe. Some enterprise customers particularly resellers and mobile service providers need wholesale and mobile backhaul network transport services. Carriers provide these specialized services as well. The term "Ethernet," of course, as we have covered in detail, refers to IEEE 802.3 standard that now covers both layer 1 and 2 of the OSI seven-layer model and lastly, the term "Network" is a telecommunications network, which increasingly refers to data networks, that allows end stations to exchange data. Putting the first two terms together "Carrier Ethernet" defines standard technology agnostic service definitions that use Ethernet-based interfaces between customers and carriers. And, the term "Carrier Ethernet network" defines carrier class data network that utilizes Ethernet technology to provide services defined by the term "Carrier Ethernet." Simply stated CENs are the data networks that extend the reach of Ethernet from LAN to MAN, RAN, and WAN so that services can be provided by using Ethernet-based interfaces.

The carriers were, understandably, concerned with migrating services from local to wide area too quickly using the Ethernet technology because they had experienced and invested too much money in many technologies that came and went. This migration was also constrained by the inventory of devices that could be provisioned and concerns about quality. Moving too fast could mean loss of money, and moving too slow would mean loss of customers. However, as described in previous chapters and sections, the evolution of Ethernet has allayed these concerns. The benefits of adopting CENs stem from the fact that today over 90% of LAN traffic is on Ethernet and extending the same technology to CENs results in reduced total cost of ownership (TCO) while increasing bandwidth, supporting QoS, reducing jitter and latency, and these are the requirements of the new video-centric and time critical applications that customers are increasingly using. The reduction in TCO is due to the fact that services are provided on converged data networks with economy of scale and reduction in operating costs by elimination of multiple technology-based networks. Also, extending Ethernet from LAN to CENs using Ethernet as an interface means less training costs to customers because they are already familiar with Ethernet in their LANs.

Carrier Ethernet and associated CENs are covered in more detail in the next chapter.

3.5 Chapter Summary

This Chapter expanded on the conclusions of first two chapters about data network and its architecture especially based on OSI seven-layer model, and delved deeper in to layer 1 and 2 of the OSI seven-layer model. The chapter described the Ethernet protocol and its role in layer 2 and its dependence on layer 1. The chapter then traced the Ethernet evolution from shared coaxial cable medium–based protocol to hub-based protocol then to bridge-based protocol, and finally to switch-based protocol. This chapter also covered the additions of Spanning Tree Protocol to provide reliability and VLAN protocol for segmenting broadcast domains as well as to provide quality-of-service functions in the Ethernet protocol. Impact of fiber optics on the Ethernet protocol, to achieve higher bandwidths over longer distances, was also traced in this chapter. This was critical for the emergence of Ethernet from LAN to MAN, and then the support of Ethernet over DWDM allowed its usage in RAN as well as WAN at layer 1 of the OSI seven-layer model. This was also covered in the chapter.

This chapter also provided implementation details of the Ethernet technology in hardware at chip and device levels. This information is useful because it is the hardware that forms the data networks over which bits move from source to destination and on this movement of bits, ride the services needed by customers.

Finally, the chapter introduced CENs on which services that leverage the Ethernet technology are provided. These services are defined by a marketing term called "Carrier Ethernet," and that is the topic of the next chapter.

4

CARRIER ETHERNET NETWORKS

What helps people, helps business.
Remarks by Leo Burnett, Advertising Executive and Founder of Leo Burnett
Worldwide

Basic research is generally driven by curiosity and sometimes leads to applied research. Applied research almost always fuels development of technology. The term "applied" implies that technology resulting from applied research has application in some product or service offering in the market. Technology without application in market is like sea water. You can swim in it all you want for fun, but you cannot drink it to survive. Market for a product or a service, on the other hand, is driven by tangible benefits to both consumers and providers of products and services. The evolution of the Ethernet technology was driven by research in the fields of protocol development, fiber optics, lasers, photodetectors, semiconductors, and microprocessors to name a few. This evolution of the Ethernet technology provided tangible benefits, first in LANs and then in wider areas outside of LANs, as we discussed in previous chapters. These benefits included reduced TCO, higher bandwidths, longer distances, low frame delay (latency), and low frame delay variation (jitter). These benefits are necessary conditions but not sufficient conditions for carriers or service providers to go to market with products and services using Ethernet technology. The other five conditions needed for going to market are standardized services, scalability, reliability, quality of service, and service management. Some of these requirements were addressed by various IEEE specifications. With that, the only gap that remained was to define the products and services leveraging the Ethernet technology that could be offered beyond LANs. And to fill this gap, industries including carriers, equipment manufacturers, cable MSOs, semiconductor manufacturers, systems integrators, software companies, testing companies, research organizations, and universities came together and formed a forum called Metro Ethernet Forum (MEF) in 2001. The name was changed in 2015 to "MEF Forum." This forum coined a marketing term called carrier Ethernet (CE) to define and standardize a group of services that leverages the Ethernet technology. Their initial focus was MAN but now the scope has expanded to include access networks, RAN, WAN, and even global networks as well. These networks that support services defined by carrier Ethernet are known as carrier Ethernet networks (CENs). In short, CE to a customer or end-user means ubiquitous, standardized, and carrier grade services that are scalable, reliable, support quality of service and are managed and CEN to a service provider means a network beyond LAN where services defined by CE are provided to end users.

This chapter will first cover definitions associated with carrier Ethernet. Then, it will cover types of CE services, defined by attributes and parameters that are offered on the carrier Ethernet networks (CENs). This chapter will also include traffic engineering and operations, administration, and management of CENs. Finally, this chapter will set the stage for peering carrier Ethernet networks.

4.1 Carrier Ethernet–Related Terminology and Architecture

For a carrier to offer carrier grade services, it is essential to have scalability, reliability, quality of service, and service management. These were addressed by various IEEE standards[34] and ITU-T[35] standard shown in Table 4.1 below.

Many of these standards were covered in Chapter 3 particularly IEEE 802.1D related to STP and IEEE 802.1w amendment related to RSTP which reduced the failover time to 10 s from about 50 s for STP. Although this was a big improvement, it was still high. The IEEE 802.1Qay further reduced the failover time to about 50 ms from 10 s that was in RSTP. Chapter 3 also described VLAN and associated VLAN tags specified in IEEE 802.1Q standard. Support for link aggregation (LAG) was originally in IEEE 802.3ad; however, it has since then been moved into its own standard IEEE 802.1AX in 2008 and the current version of that specification is IEEE 802.1AX-2014. In Chapter 3, we also covered support for 10 Gbps in IEEE 802.3ae and 40 and 100 Gbps bandwidths in IEEE 802.3ba.

The amendments IEEE 802.1ad and IEEE 802.1ah which were merged with IEEE 802.1Q shown in Table 4.1 provide for clear and secure demarcation of responsibilities of a customer and a service provider as far as devices and networks are concerned. These standards also cover the scalability of Ethernet networks to wider areas. IEEE 802.1Qay covers Provider Backbone Bridging–Traffic Engineering allowing a provider to manage traffic involving queuing, scheduling, and policing of the Ethernet frames. This is needed because different applications, users, and data flows in

Table 4.1 Standards for Carrier Grade Ethernet Services

Standard	Year(s) of Publication	Objectives
IEEE 802.1D	1990, 1998 and 2004	Original publication (802.1D-1990)
IEEE 802.1d (amendment)	1998	Reliability (STP). Merged in IEEE 802.1D-1998
IEEE 802.3ad (amendment)	2000	Reliability (support for LAG). Moved to IEEE 802.1AX-2014.
IEEE 802.1w (amendment)	2004	Reliability (RSTP)
IEEE 802.1Q	2005, 2011 and 2014	Reliability and CoS (included RSTP, MSTP, and VLAN)
IEEE 802.1ad (amendment)	2006	Scalability (provider bridging also known as Q-in-Q or stacked VLANs). Merged in IEEE 802.1Q
IEEE 802.1ah (amendment)	2008	Scalability (provider backbone bridging also known as MAC-in-MAC, for demarcation of customer and provider). Merged in IEEE 802.1Q
IEEE 802.1Qay	2009	Provider backbone bridging—traffic engineering
IEEE 802.3ae (amendment)	2002	Speed and distance (10 Gbps)
IEEE 802.3ba (amendment)	2010	Speed and distance (40 and 100 Gbps)
IEEE 802.1ag (amendment)	2007	OAM
ITU-T Y.1731	2015	OAM

a carrier Ethernet network require different priorities and performance guarantees. This process of differentiating traffic by queuing, scheduling, and policing of the Ethernet frames is referred to as traffic management. With traffic management in place, it is possible to guarantee a certain quality of service (QoS) for a given service with respect to data rate, frame delay, frame delay variation, and packet loss probability. IEEE 802.1ag and ITU-T Y.1731 are related to OAM for service management primarily fault and performance management. This will be covered in this chapter as well as in Chapter 5.

As these standards started coming out, the Metro Ethernet Forum (MEF) was formed in 2001 to define and standardize carrier grade services that leverages the Ethernet technology in wider areas beyond LANs. This forum coined a marketing term called carrier Ethernet (CE) to describe these services. Originally MEFs focus was limited to MANs and most of the MEF's steep growth, from about 2006 to now, is related to 10 Gbps introduced in IEEE 802.3ae standard. Now with the IEEE 802.3ba standard supporting 40 and 100 Gbps and also implementation of Ethernet over DWDM has increased the bandwidth and distance considerably to cover RAN, WAN, and global networks and with the start of certification tests for 100 Gbps starting within the last quarter of 2015, MEF has now expanded its focus from access networks to global networks and every network in between including MAN, RAN, and WAN. That is the main reason why MEF has replaced the term Metro Ethernet Network in its earlier standards with carrier Ethernet network (CEN) to describe the networks that support services defined by CE. It is interesting to note that MEF changed its name recently to MEF Forum in 2015, they probably should have changed it from Metro Ethernet Forum to Carrier Ethernet Forum (CEF Forum) to reflect the growth in the distance covered by Ethernet from MAN to RAN, WAN, and even global networks!

Goal of MEF was to standardize service definitions with attributes and parameters, but its goal did not include specifying how these services are provided by service providers on their CENs. This implementation was left to the service providers. Standardization of service definitions was necessary so that customers could compare service offerings from various providers and also to facilitate SLAs between customers and service providers. There are over 50 MEF specifications[36] that facilitate various aspects of service standardization efforts. Table 4.2 below shows only some of the MEF specifications that are more relevant to this chapter. Specifications related to peering of CENs will be covered in next chapter.

Table 4.2 Sample of MEF Standards for Carrier Ethernet

MEF Specification	Year	Objectives	Notes
1	2003	Ethernet service model	This specification is included just for historical reasons to indicate the first specification from MEF. It has been irrelevant for a long time. It was superseded by MEF 10 many years ago.
2	2004	Ethernet service protection framework	
6	2004	Ethernet service definitions	Superseded by MEF 6.1
9	2004	Abstract test suites for Ethernet services at UNI	
10	2004	Ethernet service Attributes	Superseded by MEF 10.1 and then by 10.2

Table 4.2 Sample of MEF Standards for Carrier Ethernet—continued

MEF Specification	Year	Objectives	Notes
14	2005	Abstract test Suite for traffic management	
6.1	2008	Ethernet service definitions	Superseded by MEF 6.2
10.2	2009	Ethernet service Attributes	Superseded by MEF 10.3
10.2.1	2011	Ethernet service performance Attributes	Amendment to MEF 10.2
6.1.1	2012	Layer 2 control protocol (L2CP)	Amendment to MEF 6.1
10.3	2013	Ethernet service Attributes	
30.1	2013	Service OAM fault management implementation	
6.2	2014	Ethernet EVC service definitions	
35.1	2015	Service OAM performance Monitoring Implementation agreement	

It can be noticed from Table 4.2 that these specifications have introduced new terms like UNI and EVC. It is important to understand these terms before going into services defined by carrier Ethernet. Fig. 4.1 shows the topology of a point-to-point service provided by a service provider to a customer to connect site A with another geographically distributed site Z so that the end stations in these two locations are made part of the same LAN thus enabling the end stations to exchange data. For discussion purposes, it is assumed that all links are full duplex.

End stations in site A are connected through a shared medium or a hub. This in turn is connected to port 1 on the customer's Ethernet switch known as customer edge (CE) on the access link. Customer configures the switch to direct all frames from port 1 to port 4 which is then connected to the port 1 on another Ethernet switch known as network termination equipment (NTE), also known as network interface device (NID), provided by the service provider and located at the customer premises. Port 1 on this NTE is the demarcation point between customer and the service provider as per IEEE 802.1ad. This NTE in general has many ports but in this case as shown in Fig. 4.1, only one port is provided to the customer. NTE then is configured by the service provider to direct all frames from port 1 to port 4. This port is then connected by a local loop to an Ethernet network switch in the CO of the service provider. Common practice is to connect NTE to an Ethernet multiplexer (EMUX) so that traffic from various customers in that location or building can

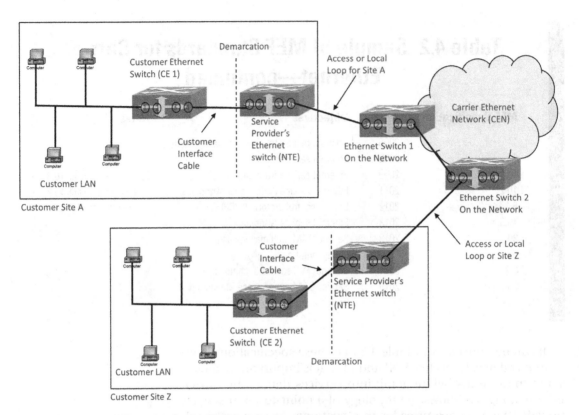

Figure 4.1 Carrier Ethernet network topology for a point-to-point service.

be multiplexed on the one local loop and sent to CO. The frames from port 1 on the network switch 1 are directed to port 4 which is connected to another Ethernet switch 2 on the carrier's network, from this switch, the frames are sent to an NTE at site Z on the local loop and from that NTE to customer's CE 2 and finally to the hub that connects all end stations in site Z. Ethernet frames from site Z to site A just follow the path in reverse direction.

The frames between site A and Z can follow a path that is essentially connectionless based on datagram approach or can be predetermined based on connection-oriented virtual connection approach. The comparison of datagram and virtual connection was already covered in Chapter 2, Section 2.1 and summarized in Table 2.1. Essentially, in connectionless mode, each frame includes complete addressing information. Each frame is labeled with a destination address and source address. The frames in this connectionless mode are switched individually, sometimes resulting in different paths. The payload in the frame carries upper layer packets and are delivered to upper layer at the destination. This precludes the need for a dedicated path to help the frame find its way to its destination, but means that much more information is needed to be looked up in content-addressable memory in the switch for processing, and therefore, the switch has to do much more work in processing every frame. In the connection-oriented approach, on the other hand, configurations of NTE and various network switches are needed to set up path of a virtual connection. This initial configuration represents additional work before any frame is transferred to establish communication, but once the path is set up, the switch has to deal with less information during operation. This path that is set up in a connection-oriented system and provided to the end-user is known as a virtual connection.

Historically Ethernet has been a connectionless technology by design. In classic LAN environments, the connectionless capabilities of Ethernet MAC bridging and CSMA/CD provided considerable flexibility, simplicity, and economy in networking latency-insensitive traffic within a single, well-bounded administrative domain. However, to bring the cost and flexibility benefits of Ethernet into the public network, the industry has made enormous modifications, enhancements, and extensions to classic Ethernet protocols. Most of these enhancements have focused on extending the connectionless classic Ethernet protocol to a service provider environment. This has brought about new capabilities in scalability, reliability, speed, distance, and service management and made the Ethernet services carrier grade. In addition, equipment vendors have enhanced enterprise-class Ethernet hardware and software platforms to provide a number of carrier-grade features in the evolution toward CEN.

The next step forward was to make Ethernet connection oriented. Connection-oriented Ethernet is optimized for aggregation infrastructure and supporting various service types under CE. The three essential functions that make Ethernet connection oriented are:

- Predetermined virtual connection paths
- Resource reservation and admission control
- Per-connection traffic engineering and traffic management

The ability to predetermine the virtual connection path through the Ethernet network is fundamental to making Ethernet connection oriented. The virtual connection path is configured through a management plane application or via an embedded control plane. This ensures that all frames in the virtual connection pass over the same sets of nodes. The frames include a connection identifier as defined in IEEE 802.1Q and are negotiated between endpoints so that they are delivered in order and with error checking. After the path has been set up at each node during the connection setup phase and the route to the destination is discovered, then the address information is transferred to each node and an entry is added to the switching table in each network node through which the connection passes.

Connection-oriented Ethernet technology provides both efficient aggregation and performance guarantees, thereby completing the Ethernet revolution by accomplishing for Ethernet what SONET/SDH accomplished for TDM. As Ethernet is playing a growing role in wide area networking for 4G mobile backhaul, enterprise services, as an access technology for IP services and residential broadband services, network providers are able to deploy a single, general-purpose Ethernet aggregation and transport environment for all these applications. This new infrastructure must deliver the robustness and performance guarantees of layer 1 SONET/SDH/OTN encapsulation, along with the aggregation and statistical multiplexing economies of native Ethernet. To create a general purpose infrastructure, Ethernet has evolved beyond simple connectionless networking technology into a carrier-grade connection-oriented technology.

Once the virtual connection path through the network has been explicitly identified, resource reservation and connection admission control (CAC) are the next critical function. In this function, actual bandwidth and queuing resources required for each virtual connection are reserved in all nodes along the path. This is vital to ensure the highest possible levels of performance in terms of frame loss probability, frame delay, and frame delay variation. CAC ensures that the requested resource is actually available in each node along the path.

Once the path has been determined and the resources allocated and after frames have been classified on network ingress, the traffic engineering and traffic management functions ensure that the requested connection performance is actually delivered. The following traffic management functions must be provided:

- Queuing
- Scheduling
- Policing
- Shaping

These traffic engineering functions determine whether each EVC reaps the performance benefit of connection-oriented Ethernet or is essentially connectionless from the performance point of view. We will discuss this in more detail in another section on traffic engineering.

Using the topology of Fig. 4.1, the combination of port on NTE, the EMUX, if present, and local loop right up to the port on the network switch is defined as user network interface (UNI). This is shown in Fig. 4.2. The MEF uses two criteria to categorize Ethernet services, namely interface type and connectivity. This UNI is one interface type and represents a service demarcation between end user and service provider. UNIs are billable items and have many attributes which depend on the service types with each attribute defined by one or more parameters as we will see in the next section. The other interface type is called ENNI which is used for peering CENs, and we will cover it in Chapter 5. The connectivity requirement is the other criteria in service categorization used by MEF and that includes, for example, point-to-point or multipoint-to-multipoint or rooted-multipoint connectivity. This connectivity is defined by Ethernet virtual connection (EVC).

The Ethernet virtual connection (EVC) shown in Fig. 4.2, is a logical association between two or more UNIs. EVCs are useful for describing the virtual connectivity of a service between UNI endpoints. Like UNI, EVCs also have many attributes which depend on the service types with each attribute defined by one or more parameters as we will see in next section. Some service providers bill the end-user for both UNIs and EVCs, and other providers only bill for UNIs and do not consider EVCs as billable items.

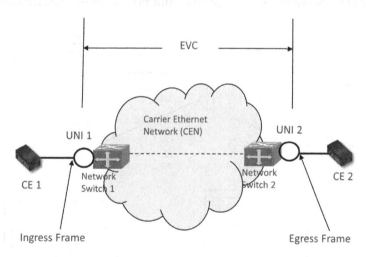

Figure 4.2 Terms associated with carrier Ethernet network.

Based on these definitions of CE, CEN, UNI, EVC and understanding of connectionless and connection-oriented Ethernet technology, MEF developed an Ethernet services framework to define Ethernet service types which are umbrella categories from which specific services can be created. This framework was developed to provide subscribers and service providers a common nomenclature when talking about the different service types and their attributes. This Ethernet services framework is discussed in detail in the next section.

4.2 Ethernet Services Framework

The schematic of the Ethernet service framework is shown in Fig. 4.3. This framework has three components, Ethernet service type, Ethernet service attributes, and Ethernet service attribute parameters. Each Ethernet service type has service attributes, and each attribute in turn has a set of parameters. The Ethernet service attribute is further subdivided into UNI attributes, EVC attributes, and EVC per UNI attributes. MEF 6.2[36] defines three service types namely (1) Point-to-Point (P2P) or Ethernet-Line or E-Line, (2) Multipoint-to-Multipoint (MP2MP) or Ethernet-LAN or E-LAN, and (3) Rooted-Multipoint or Ethernet-Tree or E-Tree. A fourth service type called Ethernet-access or E-access was added recently and is mainly used for peering CENs, and we will cover this in next chapter. The key differentiator between service types is the EVC, indicated by EVC type attribute. As mentioned before the attributes and parameters are defined in MEF 10.3.[36] More than one Ethernet service can be defined for each of these three service types. These services are differentiated by service identification at the UNI. Services using all-to-one bundling or port-based UNI are

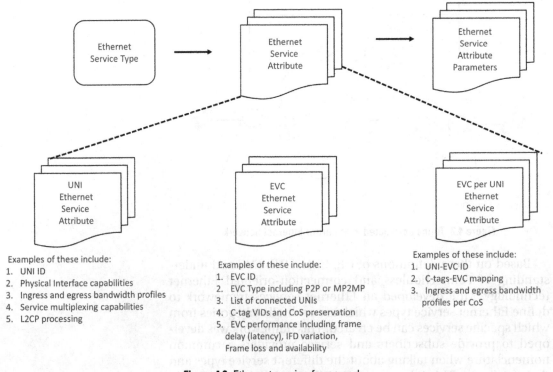

Examples of these include:
1. UNI ID
2. Physical Interface capabilities
3. Ingress and egress bandwidth profiles
4. Service multiplexing capabilities
5. L2CP processing

Examples of these include:
1. EVC ID
2. EVC Type including P2P or MP2MP
3. List of connected UNIs
4. C-tag VIDs and CoS preservation
5. EVC performance including frame delay (latency), IFD variation, Frame loss and availability

Examples of these include:
1. UNI-EVC ID
2. C-tags-EVC mapping
3. Ingress and egress bandwidth profiles per CoS

Figure 4.3 Ethernet service framework.

called "Private," whereas services using CE-VLAN ID are referred to as "Virtual Private." With these high-level classification of service types, further discussion in this section is about attributes and parameters and their groupings. We will revert to service types in the next section for more details.

It should be clear by now that IEEE has classified specifications based on scalability, reliability, QoS, and service management, and MEF has translated these into attributes and parameters at UNI, EVC, and EVC per UNI. However, there is no good information on the traceability matrix to link IEEE specifications to MEF attributes. MEF requires that for creating a specific Ethernet service, a provider has to specify which service type the service belongs to by specifying service type attribute at the EVC and then specify the UNI, EVC, and EVC per UNI attributes and associated parameter values. To better understand these attributes and their functions, the service attributes are grouped in to the following major categories:

- Ethernet physical interface attribute (speed and distance related scalability attributes)
- Bandwidth profile attribute (traffic policing attribute for QoS)
- Performance attribute (frame delay, frame delay variation, and frame loss probability related attributes for QoS)
- Class of service-related attribute (Traffic queuing and scheduling attributes for QoS)
- Service frame delivery attribute (unicast, multicast, broadcast, and L2CP-related attributes)
- VLAN tag support attribute (service delivery–related attributes)
- Service multiplexing attribute (service delivery–related attributes)
- Bundling attribute (service delivery–related attributes)
- Security filters attribute (security-related attributes)

Details of these attributes and associated parameters are given below. For additional information refer to MEF 10.3.[36]

4.2.1 Ethernet Physical Interface Attribute (Scalability-Related Attributes)

The Ethernet physical interface attribute has the following parameters:

- Physical medium—defines the physical medium per the IEEE 802.3-2012 standard. This includes speed and type of interface. It must operate in full duplex mode.
- Synchronous mode—indicates if this mode is enabled or disabled. When it is "enabled," the service provider must specify the quality of the physical reference clock that is being provided. Ethernet technology, covered in Chapter 3, is inherently asynchronous because the clock speed is valid only in each

LAN segment and not end-to-end between the ingress port and egress port of the Ethernet wide area network and therefore, incoming frames are received at one data rate, buffered and multiplexed with other frames and travel over intermediate links in the network at higher data rates and delivered at yet another rate to the receiver. There is no fixed relationship between the timing, phase, or frequency of the incoming bit stream and the outgoing bit stream from the network. This is quite different from the principles of time division multiplexed (TDM) transmission technologies such as PDH, SDH, and SONET used earlier in wide area transport networks. In TDM, each stream of information to be transferred over the network is allocated a specific timeslot in the transmission system, a procedure that requires careful frequency and phase synchronization of all intermediate network nodes handling the flow of passing bits. Today, some services such as circuit-switched telephony and the storage area networks in data centers are still based on TDM technologies, but increasingly, this TDM traffic is being transported over packet-optical networks. Therefore, it becomes necessary to emulate a traditional wireline circuit over an Ethernet network and to maintain synchronization between the ingress port and egress port of the Ethernet wide area network. This attribute enables support when synchronization is needed by deriving extremely accurate frequency source from a primary reference clock (PRC). Clocks derived in this manner are said to be traceable to a PRC. The primary clock signal is propagated through the network to synchronize all necessary devices.

- MAC layer—specifies that MAC layer is supported as specified in the 802.3-2012 standard.

4.2.2 Bandwidth Profile Attribute (Traffic Policing– Related Attribute)

The MEF defines bandwidth profiles as a set of traffic parameters that can be applied at the UNI or at an EVC or applied per CoS identifier basis. A bandwidth profile is a limit on the rate at which Ethernet frames can traverse the UNI or the EVC or per CoS identifier and defines traffic that is less than the full bandwidth available at the UNI. There are two types of bandwidth profiles namely an ingress bandwidth profile and an egress bandwidth profile. An ingress bandwidth profile controls the amount of traffic entering a UNI or an EVC or a CoS identifier, whereas an egress bandwidth profile regulates the amount of traffic leaving a UNI, an EVC, or a CoS identifier. An algorithm is associated with a bandwidth profile to determine the Ethernet frame compliance in accordance with specified set of bandwidth profile–related parameters.

The bandwidth profile service attributes are as follows:

- Ingress and egress bandwidth profile per UNI
- Ingress and egress bandwidth profile per EVC
- Ingress and egress bandwidth profile per CoS identifier

Ingress bandwidth profiles enforcement takes place at the UNI based on the disposition of the Ethernet frame. The enforcement of the egress bandwidth profile is done at the network switch.

The bandwidth profile service attributes consist of the following traffic parameters:

- CIR (committed information rate)—this is the minimum guaranteed throughput that the network must deliver for the service under normal operating conditions. A service can support a CIR per VLAN on the UNI interface; however, the sum of all CIRs should not exceed the physical port speed. The CIR has an additional parameter associated with it called the committed burst size (CBS). The CBS is the size up to which subscriber traffic is allowed to burst in profile and not be discarded or shaped. The in-profile frames are those that meet the CIR and CBS parameters. The CBS may be specified in Kb or Mb. If, for example, a subscriber is allocated a 3-Mbps CIR and a 50-Kb CBS, the subscriber is guaranteed a minimum of 3 Mbps and can burst up to 50 Kb of traffic and still remains within the SLA limits. If the traffic bursts above 50 Kb, the traffic may be dropped or delayed.
- EIR (excess information rate)—the EIR specifies the rate above the CIR at which traffic is allowed into the network and that may get delivered if the network is not congested. The EIR has an additional parameter associated with it called the excess burst size (EBS). The EBS is the size up to which the traffic is allowed to burst without being discarded. The EBS can be specified in Kb or Mb, similar to CBS. A sample service may provide a 3-Mbps CIR, 50-Kb CBS, 10-Mbps EIR, and 80-Kb EBS. In this case, the following behavior occurs:
  - Traffic is less than or equal to CIR (3 Mbps)—Traffic is in profile with a guaranteed delivery. Traffic is also in profile if it bursts to CBS (50 Kb) and may be dropped or delayed if it bursts beyond 50 Kb.
  - Traffic is more than CIR (3 Mbps) and less than EIR (10 Mbps)—Traffic is out of profile. It may get delivered if the network is not congested and the burst size is less than EBS (80 Kb).
  - Traffic is more than EIR (10 Mbps)—Traffic is discarded.

4.2.3 Performance Attribute (Frame Delay, Frame Delay Variation, and Frame Loss–Related QoS Attributes)

The performance parameters indicate the service quality experienced by the subscriber. They consist of the following:

- Availability
- Frame Delay
- Frame Delay variation
- Frame Loss Probability

Availability is specified by the following service attributes not all of which are strictly MEF terminology but commonly used

- UNI service activation time—specifies the time from when the new or modified service order is placed to the time service is activated and usable. Remember that the main value proposition that an Ethernet service claims is the ability to cut down the service activation time to hours versus months with respect to the traditional telco model.
- UNI mean time to restore (MTTR)—specifies the time it takes from when the UNI is unavailable to when it is restored. Unavailability can be caused by a failure such as a fiber cut.
- EVC service activation time—specifies the time from when a new or modified service order is placed to when the service is activated and usable. The EVC service activation time begins when all UNIs are activated. For a multipoint EVC, for example, the service is considered active when all UNIs are active and operational.
- EVC availability—specifies how often the subscriber's EVC meets or exceeds the frame delay, frame delay variation, and frame loss service performance over the same measurement interval. If an EVC does not meet the performance criteria, it is considered unavailable.
- EVC (MTTR)—specifies the time from when the EVC is unavailable to when it becomes available again. Many restoration mechanisms can be used on the physical layer (L1), the MAC layer (L2), or the network layer (L3).

Frame delay is a critical parameter that significantly impacts the quality of service (QoS) for real-time applications. Frame delay has traditionally been specified in one direction as one-way delay or end-to-end delay. The delay between two sites is an accumulation of delays, starting from one UNI at one end, going through the CEN, and going through the UNI on the other end. The delay at the UNI is affected by the line rate at the UNI connection and the supported Ethernet frame size. For example, a UNI connection with 10 Mbps and 1518-byte frame size would cause 1.2 ms of transmission delay $(1518 \times 8 / 10^6)$.

The CEN itself introduces additional delays based on the network backbone speed and level of congestion. The delay performance is defined by an agreed on value, for example, the agreed value could be 95th percentile (95 percent) of the delay of successfully delivered egress frames over a time interval. In this example, a delay of 15 ms over 24 h means that over a period of 24 h, 95% of the "delivered" frames had a one-way delay of ≤15 ms. There is nothing magical about 95th percentile. In fact, MEF recommendations are >99th percentile. The delay parameter is used in the EVC performance attribute.

Frame delay variation is another parameter that affects the service quality. Frame delay variation is also known as jitter. It has a very adverse effect on real-time applications such as IP telephony. The frame delay variation parameter is used in the EVC performance attribute.

Frame loss probability indicates the percentage of Ethernet frames that are in-profile and that are not reliably delivered between UNIs over a time interval. On a P2P EVC, for example, if 100 frames have been sent from a UNI on one end and 90 frames that are in profile have been received on the other end, the loss would be $(100-90)/100 = 10\%$. Loss can have adverse effects, depending on the application. Applications such as email and HTTP web browser requests can tolerate more loss than VoIP, for example. The loss parameter is used in the EVC performance attribute.

4.2.4 Class of Service-Related Attributes (Queuing and Scheduling Related to QoS)

Service providers may offer on their CENs different classes of service (CoS) to subscribers defined by various CoS identifiers. Based on a white-paper titled Metro Ethernet Services—A Technical Overview by Ralph Santitoro, published by the MEF in 2003, the CoS identifiers include the following:

- Physical port—this is the simplest form of CoS that applies to the physical port of the UNI connection. All traffic that enters and exits the port receives the same CoS. It is also the least flexible. If the subscriber has different services that require different CoS then they would be required to purchase separate physical ports for each service with different CoS and that would be very expensive.
- CE—VLAN ID–based CoS (IEEE 802.1Q Priority Code Point)—this is a very practical way of assigning CoS if the subscriber has different services on the physical port where a service is defined by a VLAN ID. We have discussed this in Chapter 3, Section 3.1, and the VLAN tag format is shown in Fig. 3.7.

The 802.1p p-bit values in the IEEE 802.1Q VLAN tag (Fig. 3.7) allows the carrier to assign up to eight different levels of priorities to the customer traffic. Generally, a service provider assigns certain performance parameters such as frame delay, frame delay variation, frame loss, and availability to each CoS level. This bundling of CoS and performance parameters together with bandwidth profile defines QoS. Ethernet switches use this CoS field to specify some basic forwarding priorities, for example, frames with priority number 7 get forwarded ahead of frames with priority number 6, and so on. This is one method that can be used to differentiate between VoIP traffic and regular traffic or between high-priority and best-effort traffic. In all practicality, service providers are unlikely to exceed three levels of priority for the sake of manageability.

- IP ToS/DiffServ—the CoS may also use additional behaviors assigned by IP ToS or DiffServ applied to frames. The IP ToS field is a 3-bit field inside the IP packet that is in Ethernet frame's payload and used to provide eight different classes of service known as IP precedence. This field is similar to the PCP field if used for basic forwarding priorities; however, it is located inside the IP header rather than the Ethernet 802.1Q tag. DiffServ has defined a more sophisticated CoS scheme than the simple forwarding priority scheme defined by IP ToS. DiffServ allows for 64 different CoS values called DiffServ code points (DSCPs). Although DiffServ gives much more flexibility to configure CoS parameters, service providers are still constrained with the issue of manageability. Beyond that, the overhead of maintaining these services and the SLAs associated with them becomes cost prohibitive. Unlike CE, VLAN CoS, IP ToS, and DiffServ require network switch to inspect IP packet header in payload of the Ethernet frame to determine ToS or DSCP value. Generally Ethernet switches support this capability. Because both IP ToS and DiffServ are applied to IP packets and not Ethernet frames, additional details are out of scope for this book.

In MEF 10.3 published in 2013, the CoS identifiers are classified as

- EVC Based—each ingress service frame mapped to a given EVC has a single class of service identifier. The class of service identifier can be determined from inspection of the content of the ingress service frame. When class of service identifier is based on EVC, all ingress data service frames mapped to the EVC must map to the same class of service at the given UNI associated with that EVC.

- PCP based—when the class of service identifier is based on the priority code point (PCP) field, the CE-VLAN CoS must determine the class of service name. PCP was described in Chapter 3 and is the field that refers to CE-VLAN CoS in the customer VLAN tag in a tagged service frame.
- Internet Protocol based—when the class of service identifier is based on Internet Protocol, the class of service identifier is determined from the DSCP for a data service frame carrying an IPv4 or an IPv6 packet.

4.2.5 Service Frame Delivery Attribute (Unicast, Multicast, Broadcast, and L2CP Related)

To ensure the full functionality of the subscriber network, it is important to have an agreement between the subscriber and the CEN carriers on which frames get carried over the network and which do not. The frames traversing the network could be data frames or control frames. Some Ethernet services support delivery of all types of Ethernet protocol data units (PDUs), others may not support delivery of all types of frames. The EVC service attribute can define whether a particular frame is discarded, delivered unconditionally, or delivered conditionally for each ordered UNI pair.

Data frame processing—different possibilities of the Ethernet data frames are as follows:

- Unicast frames—these are frames that have a specified destination MAC address. If the destination MAC address is known by the network, the frame gets delivered to the exact destination. If the MAC address is unknown, the LAN behavior is to flood the frame within the particular VLAN.
- Multicast frames—these are frames that are transmitted to a select group of destinations. This would be any frame with the least significant bit of the destination address set to 1, except for broadcast, where all bits of the MAC destination address are set to 1.
- Broadcast frames—IEEE 802.3 defines the broadcast address as a destination MAC address of FF-FF-FF-FF-FF-FF.

Layer 2 control processing—different L2CP frames are needed for specific applications. For example, BPDU packets are needed for STP. The provider might decide to tunnel or discard these frames over the EVC, depending on the service. The following is a list of currently standardized L2 protocols that can flow over an EVC:

- IEEE 802.3x MAC control frames—IEEE 802.3x is an XON/XOFF flow-control mechanism that lets an Ethernet interface send a PAUSE frame in case of traffic congestion on the egress of the Ethernet switch. The 802.3x MAC control frames have destination address 01-80-C2-00-00-01. PAUSE frames are not processed in a carrier Ethernet service. They are ignored.

- Link aggregation control protocol (LACP)—this protocol allows the dynamic bundling of multiple Ethernet interfaces between two switches to form an aggregate bigger pipe. The destination MAC address for these control frames is 01-80-C2-00-00-02.
- IEEE 802.1x port authentication—this protocol allows a user (an Ethernet port) to be authenticated into the network via a back-end server, such as a RADIUS server. The destination MAC address is 01-80-C2-00-00-03.
- Generic Attribute Registration Protocol (GARP)—the destination MAC address is 01-80-C2-00-00-2X.
- STP—the destination MAC address is 01-80-C2-00-00-00.
- All-bridge multicast—the destination MAC address is 01-80-C2-00-00-10.

4.2.6 VLAN Tag Support Attribute (Service Delivery Related)

VLAN tag support provides a set of capabilities that are important for service frame delivery. Enterprise LANs are single-customer environments, meaning that the end users belong to a single organization. VLAN tags within an organization are indicative of different logical broadcast domains, such as different workgroups. CEN creates a different environment in which the Ethernet network supports multiple enterprise networks that share the same infrastructure, and in which each enterprise network can still have its own segmentation. In CENs support for different levels of VLANs and the ability to manipulate VLAN tags become very important.

Consider the example of a multitenant building in which the CE provider installs a switch in the basement that offers multiple Ethernet connections to different small offices in the building. In this case, from a carrier perspective, each customer is identified by the physical Ethernet interface port that the customer connects to. Although identifying the customer itself is easy, isolating the traffic between different customers becomes an interesting issue and requires some attention on the provider's part. Without special attention, traffic might get exchanged between different customers in the building through the basement switch. VLANs can be used to separate physical segments into many logical segments; however, this works in a single-customer environment, where the VLAN has a global meaning. In a multicustomer environment, each customer can have its own set of VLANs that overlap with

VLANs from another customer. To work in this environment, carriers are adopting a model very similar to how frame relay and ATM services have been deployed. In essence, each customer is given service identifiers which identify EVCs over which the customer's traffic travels. In the case of Ethernet, the VLAN ID given by a carrier called S-tag becomes that identifier. The carrier needs to assign to each physical port a set of VLAN IDs that are representative of the services sold to each customer. For example, Customer 1 is assigned VLAN 10, customer 2 is assigned VLAN 20, and customer 3 is assigned VLAN 30. VLANs 10, 20, and 30 are carrier-assigned VLANs that are independent of the customer's internal VLAN assignments. To make that distinction, the MEF has given the name CE-VLANs to the customer's internal VLANs and the CE–VLAN ID is called C-tag, and service provider assigned tag is called S-tag. For this schema of customer's C-tag and service provider's S-tag, there are two types of VLAN tag support. They are

- VLAN tag preservation
- VLAN tag translation/swapping

VLAN tag preservation—with VLAN tag preservation, all Ethernet frames received from the subscriber need to be carried untouched within the provider's network across the EVC. This means that the VLAN ID at the ingress of the EVC is equal to the VLAN ID on the egress. This is typical of services such as LAN extension, where the same LAN is extended between two different locations and the enterprise-internal VLAN assignments need to be preserved. Because the carrier's Ethernet switch supports multiple customers with overlapping CE-VLANs, the carrier's switch needs to be able to stack its own VLAN assignment on top of the customer's VLAN assignment to keep the separation between the traffic of different customers. This concept is called Q-in-Q stacking. With Q-in-Q, the carrier VLAN ID becomes indicative of the EVC, whereas the customer VLAN ID (CE-VLAN) is indicative of the internals of the customer network and is hidden from the carrier's network. For the service to work, the Q-in-Q function must work on a per-port basis, meaning that each customer can be tagged with a different carrier VLAN tag. This was called Q-in-Q for a while before there was a standard. After the IEEE 802.1ad standard was published, it is now known as provider bridging. Some nonstandard enterprise switches on the market can perform a double-tagging function; however, these switches can only assign same VLAN ID as a carrier ID for all ports in the switch. These types of switches work only if a single customer is serviced and the carrier wants to be able to carry the customer VLANs transparently within its network. These switches do not work when the carrier switch is servicing multiple customers, because it is impossible to differentiate between these customers using a single-carrier VLAN tag. Because of this reason, it is important that Ethernet switch support dual tagging on a per-port basis.

VLAN Tag Translation/Swapping—VLAN tag translation or swapping occurs when the VLAN tags are local to the UNI, meaning that the VLAN tag value, if it exists on one side of the EVC, is independent of the VLAN tag values on the other side. In the case where one side of the EVC supports VLAN tagging and the other side does not, the carrier removes the VLAN tag from the Ethernet frames before they are delivered to the destination. Another case is where two organizations that have merged and want to tie their LANs together, but the internal VLAN assignments of each organization do not match. The provider can offer a service where the VLANs are removed from one side of the EVC and are translated to the correct VLANs on the other side of the EVC. Without this service, the only way to join the two organizations is via IP routing which of course is at layer 3, which ignores the VLAN assignments and delivers the traffic based on IP addresses. Another example of tag translation is a scenario where different customers are given Internet connectivity to an Internet service provider (ISP). The carrier gives each customer a separate EVC. The carrier assigns its own VLAN ID to the EVC and strips the VLAN tag before handing off the traffic to the ISP. In an example where a carrier-delivering Internet connectivity to three customers, the carrier is receiving untagged frames from the customer edge devices located at each customer premises. The carrier inserts a VLAN tag 10 for all of customer 1's traffic, VLAN 20 for customer 2's traffic, and VLAN 30 for customer 3's traffic. The carrier uses the VLAN tags to separate the three customers' traffic within its own network. At the point of presence, the VLAN tags are removed from all EVCs and handed off to an ISP router, which is offering the Internet IP service.

4.2.7 Service Multiplexing Attribute (Service Delivery Related)

Service multiplexing is used to support multiple instances of EVCs on the same physical connection at a UNI. This allows the same customer to have different services on one Ethernet port instead of buying multiple ports thus saving money. It maximizes the port utilization and reduces customer's equipment cost, space, power, and cabling requirements. From operator's perspective, it simplifies adding new services by simply adding new EVC to the existing UNI and avoiding site visits to add new ports, cross connects, and patch cables.

4.2.8 Bundling Attribute (Service Delivery Related)

The Bundling service attribute enables two or more VLAN IDs to be mapped to a single EVC at a UNI. With bundling, the provider and subscriber must agree on the VLAN IDs used at the UNI and the mapping between each VLAN ID and a specific EVC. A special case of bundling is where every VLAN ID at the UNI interface maps to a single EVC. This service attribute is called all-to-one bundling.

4.2.9 Security Filters Attribute (Security Related)

Security filters are MAC access lists that the carrier uses to block certain addresses from flowing over the EVC. This could be an additional service a carrier can offer at the request of a subscriber who would like a level of protection against certain MAC addresses. MAC addresses that match a certain access list could be dropped or allowed.

This information about the major categories of attributes will be helpful in examining the attributes at UNI, EVC, and EVC per UNI shown in Tables 4.3–4.5. It is important to remember that specific services can be defined by assigning different values to parameters related to each attribute. The Ethernet service attributes and their associated parameters for UNI are shown in Table 4.3 below per MEF 10.3.[36]

Table 4.3 UNI Service Attributes

Attribute	Type of Parameter Value
UNI ID	A non-null string no greater than 45 characters
Physical layer	A standard Ethernet PHY per IEEE 802.3-2012 but excluding 1000BASE-PX-D and 1000BASE-PX-U for each physical link implementing the UNI. This specifies both the type of physical interface and speed. The physical layer must operate in full duplex mode.
Synchronous mode	Enabled or disabled for each physical link implementing the UNI.
Number of links	≥1
UNI resiliency attribute	None or link aggregation
Service frame format (MAC Layer)	Per IEEE 802.3-2012
UNI maximum service frame size	≥1522
Service multiplexing	Enabled or disabled. If enabled, all-to-one bundling must be disabled.
CE-VLAN ID	An integer in 1, 2, …, 4094. Note VLAN ID 0 is reserved for priority tagged service frames and 4095 is reserved for special purposes.
CE-VLAN ID/EVC map	Mapping table of CE-VLAN IDs to EVC (mapped as per Section 9.10 of MEF 10.3)

Continued

Table 4.3 UNI Service Attributes—continued

Attribute	Type of Parameter Value
Maximum number of EVCs	Integer greater than or equal to 1
Bundling	Enabled or disabled. Must be disabled if all-to-one bundling is enabled and enabled if all-to-one bundling is disabled.
All-to-one bundling	Enabled or disabled. If enabled, service multiplexing and bundling must be disabled. Must be disabled if bundling is enabled.
Ingress bandwidth profile per UNI	No (no bandwidth profile per UNI is set) or set the traffic parameters CIR, CBS, EIR, EBS, CM, CF as defined in Section 12.1 of MEF 10.3.
Egress bandwidth profile per UNI	No (no bandwidth profile per UNI is set) or set the traffic parameters CIR, CBS, EIR, EBS as defined in Section 12.1 of MEF 10.3.
Link OAM	Enabled or disabled
UNI MEG	Enabled or disabled
E-LMI	Enabled or disabled
Layer 2 control protocols processing (L2CP)	Process, discard, or pass to EVC the following L2CP service frames per MEF 6.1.1:
	1. IEEE 802.3x MAC control
	2. Link aggregation control protocol (LACP)
	3. IEEE 802.1x port authentication
	4. GARP
	5. STP
	6. Protocols multicast to all bridges in a bridged LAN

Reproduced with permission of the Metro Ethernet Forum.

The generic EVC attributes and parameters are shown in Table 4.4 below.

Table 4.4 EVC Service Attributes

Attribute	Type of Parameter Value
EVC type	Point-to-point, multipoint-to-multipoint, or rooted-multipoint
EVC ID	A non-null string no greater than 45 characters
UNI list	A List of <UNI ID, UNI Role> pairs
Maximum number of UNIs	Integer. 2 if EVC type is point-to-point, otherwise greater than or equal to 3.
Unicast data service frame delivery	Deliver unconditionally or deliver conditionally.
Multicast data service frame delivery	Deliver unconditionally or deliver conditionally.
Broadcast data service frame Delivery	Deliver unconditionally or deliver conditionally.
CE-VLAN ID preservation	Enabled or disabled

Table 4.4 EVC Service Attributes—continued

Attribute	Type of Parameter Value
CE-VLAN CoS preservation	Enabled or disabled
EVC performance	Performance objectives and parameters including CoS identifier, frame Delay, frame jitter, frame loss as described in Section 8.8 of MEF 10.3.
EVC maximum service frame size	≥1522
Layer 2 control protocols processing (L2CP)	Process, discard, or pass to EVC the following L2CP service frames per MEF 6.1.1: 1. IEEE 802.3x MAC control 2. Link aggregation control protocol (LACP) 3. IEEE 802.1x port authentication 4. GARP 5. STP 6. Protocols multicast to all bridges in a bridged LAN

Reproduced with permission of the Metro Ethernet Forum.

The generic EVC per UNI service attributes, applicable at the end of EVC interfacing with a UNI, are listed in Table 4.5 along with their possible parameter values.

Table 4.5 EVC per UNI Service Attributes

Attribute	Type of Parameter Value
UNI EVC ID	A string formed by the concatenation of the UNI ID and the EVC ID
Class of service identifiers	Basis and values as described in Section 10.2 of MEF 10.3
Color identifier (Section 10.3)	Basis and values as described in Section 10.3 of MEF 10.3.
Egress equivalence class identifier	Basis and values as described in Section 10.4 of MEF 10.3.
Ingress bandwidth profile per EVC	No or parameters as defined in Section 12.1 of MEF 10.3
Egress bandwidth profile per EVC	No or parameters as defined in Section 12.1 for each EVC as per MEF 10.3.
Egress bandwidth profile per egress equivalence class identifier	No or parameters as defined in Section 12.1 for each egress equivalence Class 32
Source MAC address limit	Enabled or disabled. If enabled, a positive integer and a positive time interval.
Test MEG	Enabled or disabled
Subscriber MEG MIP	Enabled or disabled

Reproduced with permission of the Metro Ethernet Forum.

A broad range of Ethernet services can be constructed by applying different parameter values to the attributes for UNI, EVC, and EVC per UNI given in Tables 4.3–4.5.

This background information of the Ethernet services framework and the associated attributes and parameters will help in understanding CE service types in details covered in the next section.

4.3 Services Defined by Carrier Ethernet

The CE service types cover E-Line, E-LAN, E-Tree and E-Access. E-Line, E-LAN, and E-Tree service types encompass all services that interconnect UNIs and represent point-to-point, multipoint-to-multipoint, and rooted-multipoint connectivity, respectively. The E-Access service type encompasses all services that interconnect an ENNI with at least one UNI, and this will be covered in the next chapter because this service type is key for peering CENs of different providers or carriers. Within each Ethernet service type, the MEF defines Ethernet services based on whether frames are mapped to EVCs based on the port or mapped to EVCs based on the C-VLAN. In the port-based or private services case, there can be no service multiplexing, and in the VLAN-based or virtual private services case, there may or may not be service multiplexing. One of the benefits of a VLAN-based service is the ability to multiplex multiple services onto a single UNI saving the cost of an additional Ethernet port on an Ethernet service provider's NTE. These CE service types are shown in Table 4.6 below.

Table 4.6 Carrier Ethernet Service Types

Service Type	Port-Based Services	VLAN-Based Services
E-Line (point-to-point EVC) UNI to UNI	Ethernet Private Line (EP-Line)	Ethernet Virtual Private Line (EVP-Line)
E-LAN (multipoint-to-multipoint EVC) UNIs to UNIs	Ethernet Private LAN (EP-LAN)	Ethernet Virtual Private LAN (EVP-LAN)
E-Tree (rooted-multipoint EVC) root UNI(s) to leaf UNIs	Ethernet Private Tree (EP-Tree)	Ethernet Virtual Private Tree (EVP-Tree)
E-Access (point-to-point OVC) UNI(s) to ENNI	Access Ethernet Private Line (Access EPL)	Access Ethernet Virtual Private Line (Access EVPL)

4.3.1 EP-Line Service

Fig. 4.4 shows an example of physical architecture of an EP-Line service type where two sites A and Z of a subscriber are connected through a CEN by a provider. Subscriber's internal network or LAN at each site connects to a customer edge device (CE), and these CEs in turn connect to NTEs provided by the service provider. These NTEs in turn connect to network switches using the local loop. The logical network architecture is shown in Fig. 4.5. This is an example of point-to-point EVC where all frames from both UNIs map to this one EVC. Because all frames map to EVC, it is also known as port-based EP-Line or all-to-one bundling. The port-based E-Line service is generally used to connect two LANs or two VLAN-based subscriber networks which implement C-tags.

In case of connecting a given subscriber's two LANs, the ingress frames may be untagged or may have priority tag. On egress, tags are untagged. The priority tag is used only at ingress for CoS treatment and then removed. This service can also connect two VLANs in which case the frames may be untagged, priority tagged, or

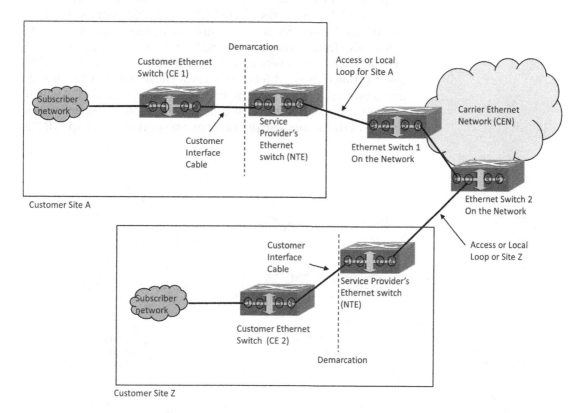

Figure 4.4 Physical architecture of an EP-Line service.

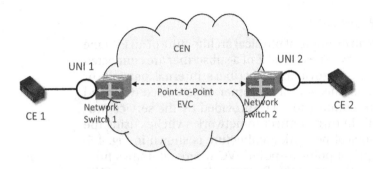

Figure 4.5 Logical view of an EPL service.

C-tagged at ingress and may be untagged or C-tagged at egress. Because there is only one VLAN at UNI 1 and UNI 2, all traffic still flows through the same EVC.

4.3.2 EVP-Line (EVPL) Service

In case of EVP-Line service in conjunction with a subscriber's multiple VLAN networks where VLAN tags are specified by IEEE 802.1Q-2011 specification, the ingress frames have C-tags. At egress, the frames may or may not have same C-tags as was at ingress. This type of service is called bundled EVPL service. This case is shown in Fig. 4.6 below. Here a subscriber has VLAN 1, 2, and 3 on both ends. These are connected to CE 1 and CE 2 which assign C-tags 1001, 1002, and 1003. These customer edge devices multiplex the frames and put them on one port on UNI 1 and UNI 2

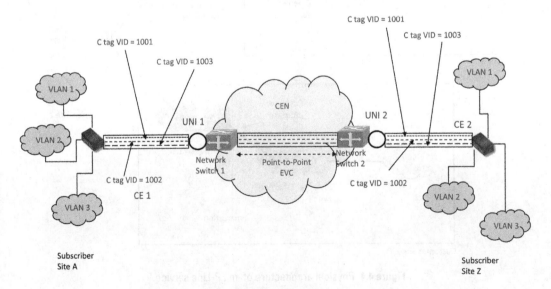

Figure 4.6 Logical view of EVPL (Bundled EVPL) service.

at site A and Z respectively. All these frames traverse through CEN while retaining their C-tags and at the receiving customer edge device, they are forwarded to correct VLANs based on these C-tags. In the CEN, these frames are encapsulated by a provider S-tag at the ingress network switch to keep traffic separated from other traffic and also to use the S-tag for establishing the connection-oriented path. This S-tag is stripped at the egress switch.

In a special case, when out of multiple subscriber VLANs, only one VLAN on both ends are connected by the EVC, then this service is called nonbundled EVC. Here C-tag can be retained while traversing the EVC and be the same at both ends or can be translated from ingress UNI to egress UNI.

4.3.3 EP-LAN

E-LAN type of service is useful when a customer has more than two locations that are needed to be connected. In case of E-LAN also, the service could be port based or VLAN based. The port-based service known as EP-LAN, service is used to connect a group of customer LANs or a group of VLANs as long as all the traffic is bundled to one EVC that is all-to-one bundled case. This service can have untagged or priority tagged or C-tags at ingress and can have untagged or C-tagged frames at egress. The physical architecture of EP-LAN service is shown in Fig. 4.7. Here a customer

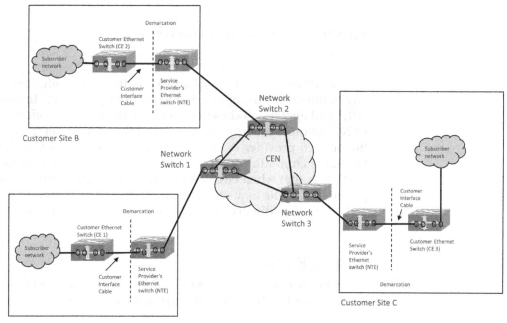

Figure 4.7 Physical architecture of an EP-LAN service.

has three locations designated by A, B, and C. Each location has a LAN that is connected by the EP-LAN service such that frames can go from site A to both B and C and similarly from site B to both A and C and from site C to site A and B. In other words, all these sites become part of one extended LAN.

From Fig. 4.7, it is clear that, for example, frames from site A can go to site C by two different paths. However, in a connection-oriented approach, the path and backup path in case of failure of the main path are specified and not determined by STP (or RSTP) protocol. The logical implementation of EP-LAN is shown in Fig. 4.8.

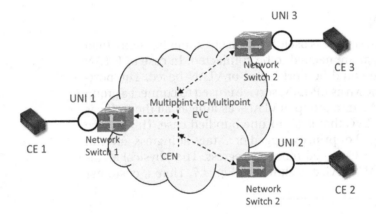

Figure 4.8 Logical view of an EP-LAN service.

The advantage of EP-LAN service is that it provides any-to-any connectivity using multipoint-to-multipoint EVC between UNIs and enables all sites to communicate with each other over CEN. Customer orders one single service to connect all sites, and the service provider does all the configuration and switching to enable communication between all sites. Another benefit of EP-LAN is that when customer wants to add a new site, service provider adds the new site to this multipoint-to-multipoint EVC, and other sites discover this new site and start communicating with it.

4.3.4 EVP-LAN

When there are one or multiple VLANs at each site and need to be connected, then EVP-LAN is the way to go. Here the C-tags at

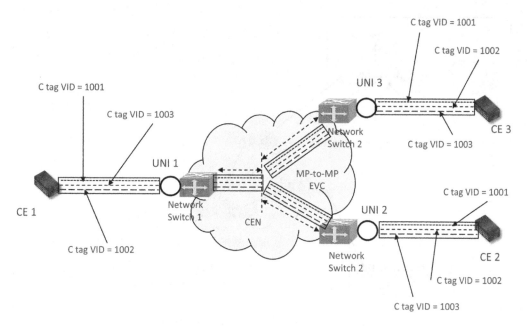

Figure 4.9 Logical view of an EVP-LAN service.

UNIs must match. The logical architecture in Fig. 4.9 shows that the same set of C-tags determine admission at each UNI in an EVP-LAN service.

The advantage of EVP-LAN is that it allows service multiplexing at UNIs.

4.3.5 EP-Tree

There are cases where branch sites or leaf sites only need to communicate with a hub or root site for example with headquarter of a company. There is no need for these leaf sites to communicate with each other directly. For such cases, EP-Tree is ideal because it can connect multiple UNIs on one EVC. The EP-Tree service is just like the EP-LAN service, but here some UNIs are prevented from forwarding frames to other UNIs. Each UNI is declared a root or a leaf UNI, and a leaf UNI can forward frames only to a root UNI. A root UNI can forward frames to all leaf and other root UNIs. Fig. 4.10 shows the physical architecture of EP-Tree service implementation.

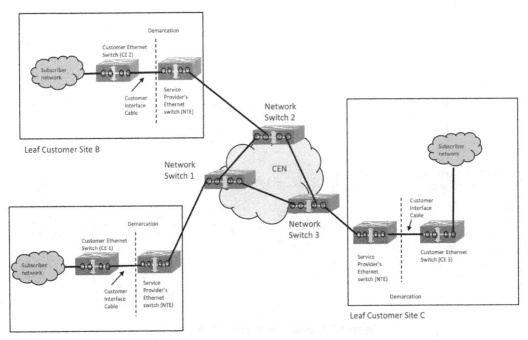

Figure 4.10 Physical architecture of an EP-Tree service.

Service similar to EP-Tree service could be achieved by having multiple point-to-point EVCs in an EP-Line implementation but if the provider charged for each UNI as well as each EVC, then the customer would end up paying more as compared to EP-Tree service where there is only one rooted-multipoint EVC as shown in Fig. 4.11.

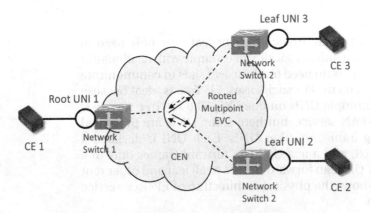

Figure 4.11 Logical view of an EP-Tree service.

4.3.6 EVP-Tree

As in case of EVP-LAN service shown in Fig. 4.9, when there are one or multiple VLANs at each site and need to be connected in a root-leaf arrangement then EVP-Tree is the way to go. Here the C-tags at UNIs must match and the same set of C-tags determine admission at each UNI in an EVP-Tree service. The advantage of EVP-Tree service is that it allows service multiplexing at UNIs.

Applying attributes and parameters of these six service types namely EP-Line (EPL), EVP-Line (EVPL), EP-LAN, EVP-LAN, EP-Tree, and EVP-Tree, many specific applications or services including site-to-site connectivity, layer 2 VPN, connection to Internet, cloud services, mobile backhaul, VoIP services, Video conferencing services, IPTV services, interactive gaming services, streaming media services, texting or messaging services, telepresence services, point of sales services, CCTV services, trading services, cyber-physical system-related services, and Internet of things (IoT)–related services can be provided. We will discuss these applications in Chapter 8 specifically devoted to applications that leverage CENs. Because these applications require an understanding of Quality of Service (QoS), service traffic engineering, and service OA&M, these are described in next two sections.

4.4 Quality of Service by Traffic Engineering of Carrier Ethernet Services

An SLA is the entire agreement between a service provider and a customer. An SLA consists of many service level objectives (SLOs) and are key elements of an SLA. Each SLO is a specific measurable characteristic such as availability, throughput, billing frequency, response time, or quality of service (QoS). SLOs are agreed as a means of measuring the performance of the service provider and are outlined as a way of avoiding misunderstanding and resultant disputes between the two parties. QoS was recognized as a key performance SLO by the industry and a pre-condition for carrier Ethernet and CEN to be carrier grade. QoS is needed because CEN capacity is not always sufficient especially for real-time services like VoIP and IP-TV. If the networks had unlimited capacity, then QoS would be redundant, but unfortunately that is not the case. Achievement of QoS was made possible by the IEEE 802.1Qay specification covering provider backbone bridging—Traffic Engineering. MEF defines service-level specification (SLS) as the technical specification of the service level offered by the service provider. The SLS includes QoS-related performance parameters such as frame delay, frame delay variation, frame loss, and CoS level as the class of service performance objectives (CPOs) for the service. We will cover CPOs in more detail in Chapters 5 and 8. The bundle of CPO and bandwidth profile defines the QoS related SLO portion of the overall SLA. Description in Section 4.1 has covered that traffic engineering involves queuing, scheduling, shaping, and policing of the Ethernet frames. The traffic engineering is needed because different applications, users, and data flows in a carrier Ethernet network require different priorities and performance guarantees. With traffic management in place, it is possible to guarantee a certain Quality of Service (QoS) for a given service with respect to data rate, CoS, frame delay, frame delay variation, and packet dropping probability. Generally, once a subscriber buys a service with certain CoS, a service provider assigns or bundles certain performance parameters such as frame delay, frame delay variation, frame loss, and availability to that CoS level.

QoS is achieved by traffic engineering and traffic engineering in turn is achieved by queuing, scheduling, policing, and shaping. In addition to these, QoS also includes performance attributes. As we saw in Section 4.2, queuing and scheduling are achieved by CoS-related attributes defined at EVC per UNI (Table 4.5), and policing is implemented by bandwidth profile-related attributes defined at UNI (Table 4.3) as well as EVC per UNI (Table 4.5). Performance-related attributes are defined at EVC (Table 4.4) per MEF 10.3 specification. Shaping is normally not part of SLA, and it is an internal mechanism used by the service provider to shape traffic. With this background, let us examine bandwidth profile, CoS, and performance attributes like frame delay, frame delay variation, and frame loss probability and finally shaping in some detail.

4.4.1 Bandwidth Profiles

A bandwidth profile is a set of traffic parameters that define the maximum average bandwidth available for the customer's traffic. An ingress bandwidth profile limits traffic transmitted into the network, and an egress bandwidth profile can be applied anywhere to control overload problems of multiple UNIs sending data to an egress UNI simultaneously. Frames that meet the profile are forwarded, and frames that do not meet the profile are dropped. Bandwidth profiles allow service providers to offer services to users in increments lower than what is set by the physical interface speed. Also, it provides a possibility to engineer the network and make sure that certain parts of the network are not overloaded. Bandwidth profile uses the two traffic parameters namely, rate expressed as bits per second and burst size expressed as bytes. Bandwidth profiles can be defined per UNI or per Ethernet Virtual Connection (EVC) or even per class of service and are governed by a set of parameters, the most important being CIR, CBS, EIR, EBS, and color mode. These are explained below in detail.

- Committed information rate (CIR): this defines the assured bandwidth expressed as bits per second. The CIR defines the bandwidth delivered per the SLO defined in the SLA. CIR bandwidth is assured for an EVC through traffic engineering across the network. To guarantee a CIR, bandwidth must be reserved across all network paths traversed by Ethernet frames associated with an EVC. Note that service performance metrics, such as frame delay or frame loss, are measured for traffic bandwidth conformant to the CIR. Traffic bandwidth that is not conformant to the CIR is excluded from performance measurements. This nonconformant traffic is considered excess and is eligible to be discarded in the network based on the traffic management policies for the service provider's network.

- Excess information rate (EIR): this defines extra bandwidth that may be temporarily used. It is expressed as bits per second. The EIR defines the amount of excess traffic bandwidth allowed into the network. EIR increases overall traffic throughput but with no SLOs as with CIR traffic. Traffic at bandwidths exceeding the CIR is considered excess and hence is not delivered per the SLO.

- Committed burst size (CBS) and excess burst size (EBS): these define temporary bursts of information that can be handled. CBS and EBS characterize the maximum number of bytes that can be injected into the network and still remain conformant with the CIR and EIR, respectively. For example, if a CBS is 50 Kb, then 50 Kb of additional Ethernet frames over and above the agreed on CIR value can be injected into and still be accepted by the network.

As mentioned before, ingress bandwidth profiles can be applied per UNI (all traffic regardless of VLAN tag or EVC ID) or more granularly on a per EVC basis or even based on a class of service marking such as a customer-applied VLAN priority tag basis. These are described below in some detail.

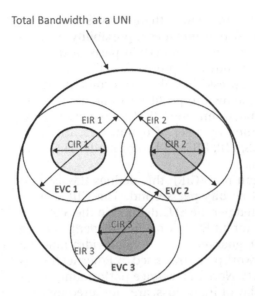

Figure 4.12 Example of UNI bandwidth divided among three EVCs.

Bandwidth profile per UNI—this type of bandwidth profile applies to the entire UNI regardless of the number of EVCs which are present. It is only useful for port-based services such as EPL which provide a single EVC at the UNI.

Bandwidth profile per EVC at a UNI—this type of bandwidth profile applies to each EVC at the UNI. When there is a single EVC at the UNI, the bandwidth profile per EVC provides comparable functionality as the bandwidth profile per UNI; however, when there are multiple EVCs at UNI, it enables the UNI bandwidth to be divided up among each EVC.

Fig. 4.12 illustrates how a UNI's bandwidth is divided up among three EVCs using the bandwidth profile per EVC. Each EVC has its specified CIR and EIR. It is clear from Fig. 4.12 that all three CIRs can be met simultaneously, but all three EIRs cannot be met simultaneously.

Bandwidth profile per EVC per CoS ID at a UNI—this type of bandwidth profile applies to Ethernet frames belonging to each CoS of an EVC. This is useful for services with an EVC at a UNI, which support multiple classes of service. Each CoS is identified by its Ethernet PCP. This bandwidth profile enables the EVC bandwidth to be partitioned by CoS. Fig. 4.13 illustrates UNI-based, EVC-based and CoS-based bandwidth profiles at the UNI.

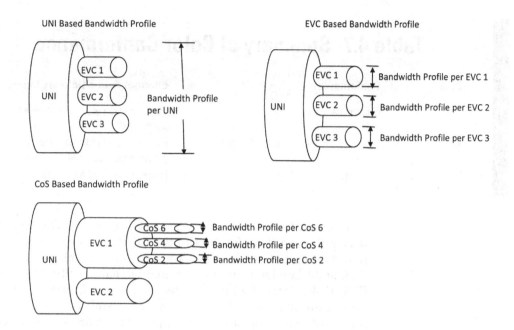

UNI Based Bandwidth Profile

EVC Based Bandwidth Profile

CoS Based Bandwidth Profile

Reproduced with permission of the Metro Ethernet Forum.

Figure 4.13 Examples of UNI-based, EVC-based and CoS-based bandwidth profiles.

Bandwidth profiles are implemented by a function called policing, and this uses a frame-coloring mechanism. In this mechanism, compliance to bandwidth profile is determined by converting two rates, namely CIR and EIR into three color mappings. This mechanism is called two-rate-three-color marker. According to MEF 10.3,[36] three color mapping for bandwidth profile compliance are

- Green: The service frame is subject to service level agreement (SLA) performance guarantees.
- Yellow: The service frame is not subject to SLA performance guarantees but will be forwarded on a "best-effort" basis. These frames are discard-eligible in the event of network congestion.
- Red: The service frame is to be discarded at the UNI by the traffic policer.

Ethernet frame three-color mappings to two rates are summarized in Table 4.7 below.

Table 4.7 Summary of Color Conformance

Color	Conformance	Ethernet Frame Delivery Expectation
Green	Conformant to CIR	Frames colored green and delivered per the SLO
Yellow	Nonconformant to CIR but conformant to EIR	Frames colored yellow and may be delivered but with no SLO assurances
Red	Nonconformant to CIR or EIR	Frames colored red and dropped

Ethernet frames are policed as they ingress at a UNI and are declared green if conformant to the CIR/CBS, declared yellow if non-conformant to the CIR/CBS and conformant to the EIR/EBS, or declared red if non-conformant to either CIR/CBS or EIR/EBS. Red-colored frames are always discarded, yellow-colored frames are discarded during times of network congestion, and green-colored frames are never supposed to be discarded. If green frames are discarded, then this counts against meeting the SLO. Color mode (CM) specifies whether the service provider should use any prior color markings of the Ethernet frames that may have been made by the subscriber's equipment or used on their LAN before those frames ingressing the UNI. If the CM is set to operate in color-blind mode, then any prior Ethernet frame coloring is ignored. If CM is set to operate in color-aware mode, then any prior Ethernet frame coloring is factored into subsequent policing decisions. Most Ethernet service offerings operate in color-blind mode because subscribers rarely, if ever, apply traffic management in their LANs that would result in any color marking of their Ethernet frames.

4.4.2 CoS

The integration of real-time and nonreal-time traffic over Ethernet requires differentiating packets from different applications and providing differentiated performance according to the needs of each application. When a network experiences congestion and delay, some packets must be dropped or delayed. This differentiation in carrier Ethernet terminology is referred to as class of service (CoS). CoS can be applied at the EVC level (same CoS for all frames transmitted over the EVC) or applied within the EVC by customer-defined priority values in the data, such as customer-defined PCP markings. The specification has provision to supports eight different CoS priorities with 0 being the lowest (best effort) and 7 being the highest priority (real-time data). The Ethernet switches have many CoS queues per port, some switches have fewer than eight, and some have more than eight but most have eight. These eight queues are serviced by a scheduler that can use three different scheduling schemes including strict, round robin, and weighted round robin for emptying the eight egress queues. The class of service settings together with the bandwidth profiles are in turn used for making QoS-related SLO in service level agreements (SLAs) between the service provider and the customers.

In addition to the bandwidth profile and CoS, various performance parameters also are part of QoS. The QoS SLO in the SLA typically also specifies SLS which includes values of frame delay, frame delay variation, frame loss probability, and availability of the subscribed service. These are described below.

4.4.3 Performance Parameters

The following performance parameters are part of the EVC performance attribute (Table 4.4):

- Availability

  Availability performance is the percentage of time within a specified time interval during which the service frame loss is small. As an example, a service provider can define the availability performance to be measured over a month and the value for the availability performance objective to be 99.9%. In a month with 30 days and no maintenance interval, this objective will allow the service to be unavailable for approximately 43 min out of the whole month. MEF 10.3[36] provides a more precise definition of availability.

- Frame delay

  Frame delay also known as latency is a critical performance attribute and has an important bearing on QoS especially for real-time applications like VoIP, IP-TV, and video conferencing. It is defined as the elapsed time delay from the transmission at the ingress UNI of the first bit of ingress service frame until the reception of the last bit of the corresponding service frame at the egress UNI. It is schematically shown in Fig. 4.14. It includes the delays encountered as a result of transmission of the service frame across the ingress and egress UNIs as well as that introduced by the CEN. Based on Fig. 4.14, frame delay is equal to A+B+C. It is measured over a time interval for a percentile of successfully delivered CIR conformant green-colored service frames. MEF 10.3[36] has additional information about frame delay.

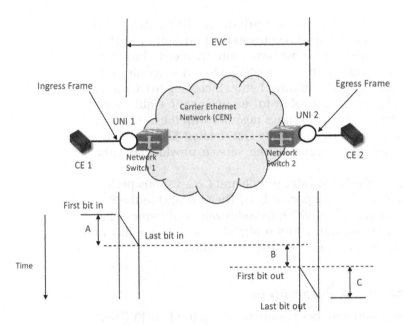

Figure 4.14 Frame delay components in CEN.

- Frame delay variation
  Frame delay variation is a critical parameter for real-time applications. These real-time applications require low frame delay variation to function properly. Frame delay variation is not a critical parameter for non-real-time applications. There are actually two ways of specifying frame delay variation. It can be specified by frame delay range which is derived from samples of frame delay by subtracting lowest frame delay from the highest frame delay in the sample data, or it can be defined by inter-frame delay variation which measures the delay variation between pairs of successive frames.
- Frame loss probability
  Frame loss probability is defined as the percentage of CIR-conformant, that is, green-colored frames not delivered between UNIs over a given measurement interval. These are also defined in MEF 10.3[36]. For example, referring to Fig. 4.14, if 100 frames were transmitted over a 5 min duration from UNI 1 to UNI 2, and of these, 99 were successfully delivered, and 1 was lost then the frame loss is 1%. The implication on QoS again depends on the application. For example, 1% loss for a real-time application like VoIP may be acceptable but 3% loss may not be acceptable.

4.4.4 Shaping

Normally, traffic shaping is not part of any subscriber SLA and therefore not specified by attributes on UNI or EVC or EVC per UNI. Shaping is not done in the CEN except maybe at the egress UNI. Shaping is a subscriber function. It is external to the CEN not internal. It is an internal mechanism, used to even out traffic flows and create fairness between users of the network resources. Traffic shaping provides a means to control by bandwidth throttling, the volume of traffic being sent out on an interface for a specified period and also to control by rate limiting the maximum rate at which the traffic is sent. Traffic shaping is a traffic management technique which delays some frames to bring them into compliance with a desired traffic profile. Traffic shaping is a form of rate limiting, as opposed to the policing of the bandwidth profiles, where excess frames are simply dropped. Traffic shaping is done by imposing additional delay on some frames such that the traffic conforms to a given bandwidth profile. A drawback with traffic shaping is increased frame delay and frame delay variation for the EVC, but the gain can be better throughout because the overall flow of frames may be improved. Instead of dropping traffic in a policer, it may be better to shape the traffic to make sure no frames are lost thus avoiding re-transmissions by higher protocol layers.

As we discussed in Chapter 3, Ethernet switch come in different shape and form, and their use is dictated by location and application. Normally smaller bandwidth and lower port density switches are used as NTE at the customer's premises as the demarcation device provided by the service provider. The traffic engineering functions are normally implemented at the NTE level and not at the network switch. Important functions of network Ethernet switches are aggregation and directing traffic.

4.5 Carrier Ethernet Network Operation, Administration, and Maintenance

Service management is part of the operation support and business support systems commonly known as OSS/BSS. This will be covered in detail in Chapter 7. Part of the operation support system (OSS) related to service assurance is operation, administration, and maintenance system, commonly known as OAM. It is another important requirement for making carrier Ethernet, a carrier grade service. Service management by OAM is important because without monitoring and remediation mechanisms, there is no way to enforce SLAs. From an OAM perspective, there are several standards that work together in a layered fashion to provide carrier

Ethernet OAM. IEEE 802.3ah defines OAM at the link level. With more of an end-to-end focus, IEEE 802.1ag defines connectivity fault management for identifying network level faults, whereas ITU-T Y.1731 adds performance management which enables SLAs to be monitored. Taking the lead from these specifications, MEF 10.3[36] defined UNI attributes as well as EVC per UNI attributes to account for service OAM (SOAM) and MEF 30.1[36] and MEF 35.1[36] specifications provided guidelines for fault management and performance management in CEN, respectively.

Recognizing the fact that Ethernet networks often encompass multiple administrative domains, IEEE 802.1, ITU-T Study Group 13, and MEF have adopted a common, multidomain SOAM reference model. The carrier Ethernet is divided into customer and service provider maintenance levels. Fig. 4.15 shows the OAM jurisdiction defined by IEEE 802.1ah, IEEE 802.1ag, and ITU-T Y.1731.

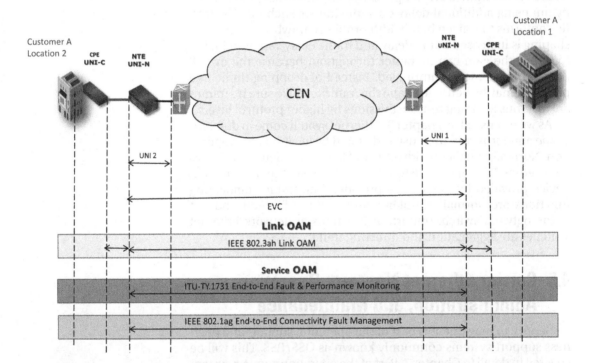

Figure 4.15 OAM jurisdiction.

As shown in Fig. 4.15, service providers have end-to-end service responsibility including NTEs and subscribers are responsible for their internal networks and CEs. In this model, a maintenance entity (ME) is defined as an entity that requires management. This is shown in Fig. 4.16.

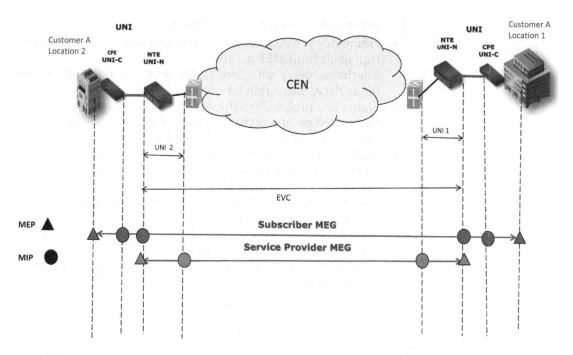

Figure 4.16 OAM framework.

An ME is essentially an association between two maintenance end points within an OAM domain, where each end point corresponds to a provisioned reference point. For example, in Fig. 4.16, the arrow between the two CEs represents a subscriber ME, and the arrow between the two NTEs represents a service provider ME. A maintenance entity group (MEG) consists of the MEs that belong to the same service inside a common OAM domain. The MEs exist within the same administrative boundary and belong to the same point-to-point or multipoint Ethernet virtual connection. For a point-to-point EVC, the MEG contains one single ME. A MEG end point (MEP) is a provisioned OAM reference point which can initiate and terminate proactive OAM frames. It can also initiate and react to diagnostic OAM frames. The MEPs are indicated by triangles in Fig. 4.16. A MEG intermediate point (MIP) is any intermediate point in a MEG that can react to some OAM frames. A MIP does not initiate OAM frames neither does it take action on the transit Ethernet traffic flows. The MIPs are indicated by circles in Fig. 4.16. The OAM functions are implemented into the node equipment or, if needed, are implemented in a stand-alone network demarcation device such as test heads, for monitoring specific functions like performance management or fault management.

Performance management includes

- Frame delay: Measurement of one-way and two-way (round-trip) delay from MEP to MEP.
- Interframe delay variation: Differences between consecutive frame delay measurements.
- Frame loss probability: The number of frames delivered at an egress UNI compared to the number of transmitted frames at an ingress UNI, over a specified time, e.g., a month.
- Availability: Downtime is measured over, for example, a year and used to calculate the availability of the service.

Fault management includes

- Continuity Check: "Heartbeat" messages are issued periodically by the MEPs and used to proactively detect loss of connection between endpoints. Continuity check is also used to detect unintended connectivity between MEGs. The continuity check is used to verify basic service connectivity and health. This is shown in Fig. 4.17.

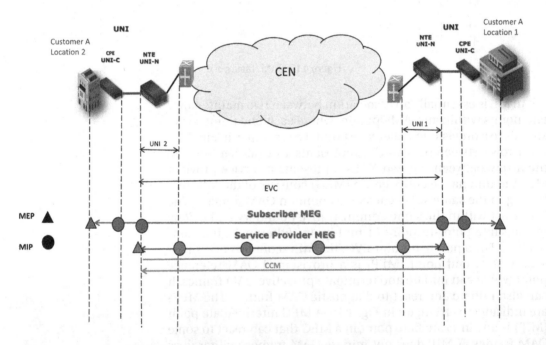

Figure 4.17 OAM fault management by continuity check message.

In case of fault detection, a MEP can communicate the fault by (1) remote defect indication signal where a downstream MEP detecting a fault will signal the condition to its upstream MEP(s). The behavior is similar to the RDI function in SDH/SONET networks or (2) alarm indication signal where a

MEP-detecting fault can send an alarm signal to its higher level MEs, thereby informing the higher level MEs of the disruption, immediately following the detection of a fault.

- Linktrace: This is an on-demand OAM function initiated in a MEP to track the path to a destination MEP. It can be thought of as a "layer 2 trace route." The procedure is similar to the loopback procedure. A MEP initiates the Linktrace by sending a Linktrace message (LTM) using a MAC DA and the appropriate MEG level for the ME. It is recommended that an LTM makes use of the highest CoS ID available, which will yield the lowest possible loss for a particular Ethernet service. Each MIP that belongs to this ME, and the remote MEP replies by generating a Linktrace Response (LTR) unicast message. In addition to sending an LTR, each MIP also simultaneously forward the LTM onward toward the remote MEP. If there are no faults in the path, Linktrace allows the originator to map the entire route to the remote MEP. If there is a fault in the path, the originator will receive LTRs from all MIPs before the fault thereby facilitating fault isolation. On the subscriber MEG, LTM PDUs should be sent with same CE-VLAN ID that maps to the monitored EVC, that way, it is guaranteed that the LTM passes the same path as the monitored EVCs service frames. It allows the transmitting node to discover connectivity data about the path. This Linktrace message is shown in Fig. 4.18 below.

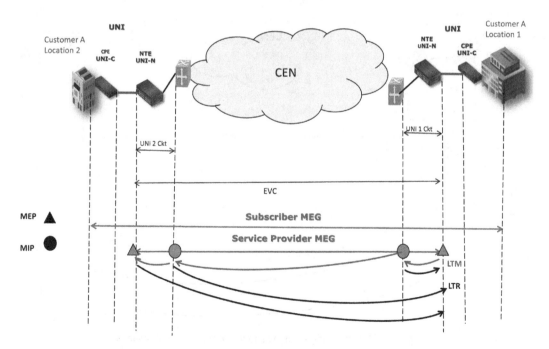

Figure 4.18 OAM fault management by Linktrace message and response.

- Loopback: This is an on-demand OAM function used to verify connectivity of a MEP with another MEP in the ME, but it can also be sent from a MEP to a MIP in the same ME. Loopback messages (LBM) are defined by IEEE 802.1Q and ITU-T Y.1731 and are used on-demand as the first step to isolate a fault, which may have been detected by CCMs. An LBM is usually sent from a MEP to a remote MIP/MEP which immediately replies with LBR (loopback reply). LBM can be thought of as "layer 2 ping." LBMs are sent as a loopback (LB) session of n (default is 3) consecutive LBMs in a predefined time interval. If an LBR is not received within a given time interval (default is 5 s), the MEP declares the remote MEP or MIP as being faulty. The LBM uses the unicast MAC DA of the destination MIP/MEP, but it can also use multicast MAC used by CCMs. It is recommended to use the highest CoS ID for an LBM, the one that yields the lowest possible loss for that Ethernet service. An LBM is usually 64 bytes long but can be extended to any value up to the MTU size. The LBR is identical to the LBM except that the MAC DA and MAC SA are swapped. This Loopback message is shown in Fig. 4.19 below.

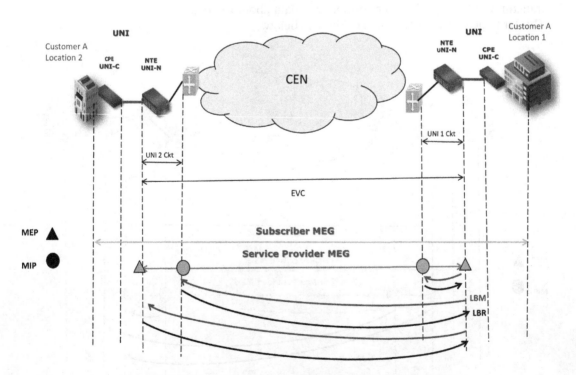

Figure 4.19 OAM fault management by Loopback message and response.

All the descriptions of CE and CENs, associated terminology, and the definitions of UNI and EVC in this chapter have been related to CENs belonging to one carrier or service provider. In the next section, the need for peering CENs is introduced to provide services to off-net subscribers.

4.6 Need for Peering Carrier Ethernet Networks

If the entire globe were to be covered by the networks of just one carrier, then things would be different, and for one thing, this book would perhaps have ended here. But the fact is that there are over 1000 carriers including tier 1, tier 2, and tier 3 carriers in the United States alone. In addition, each country in the world has multiple carriers. On the other hand, owing to global economy, many businesses have their operations around the globe, and these companies have the need to connect their branch offices with their headquarters. Similarly, mobile operators have their cell towers in other carrier's territories, and they have a need for the backhaul to connect these towers to their MSCs in other territories. In third example, there are needs to control the operations of machineries in factories from central control locations. Similarly, utility providers have the need to remotely control their grid from an operation control center. All these applications require peering of CENs. Alternate approach would be to take customer data from local access and direct the layer 2 frames over MAN and RAN on to IP layer routing on WAN and then back to RAN and MAN in another territory and finally over the local access to the customer's location at the other end. This approach of going back and forth between layer 2 and layer 3 would increase latency and reduce performance. Peering of CENs based on ENNI and using MEF's Ethernet–access services provides a way to use Ethernet technology for handling these situations and that is the topic of our next chapter.

4.7 Chapter Summary

This chapter described how communication industry came together to leverage benefits of Ethernet technology by defining and standardizing Ethernet services, making them reliable and scalable and specifying QoS and service management. This step was essential to make Ethernet-based services carrier grade. This chapter presented definitions of carrier Ethernet, carrier Ethernet networks, UNI and EVC, and described associated terminology.

The chapter also covered Ethernet service types, service attributes, and parameters and a method to deliver QoS by traffic engineering which in turn is based on queuing, scheduling, policing, shaping, and stipulating performance parameters. Queuing and scheduling are defined by class of service (CoS), policing is defined by bandwidth profile, and performance parameters are defined by frame delay (latency), frame delay variation (jitter), frame loss, and availability. Description in this chapter also included the need for SOAM functions for fault and performance monitoring to ensure that QoS is in compliance with the SLAs. All the description in this chapter was based on the assumption that CENs belonged to one carrier or service provider, and therefore, the customer was always an on-net customer. The chapter provided a transition to cases where customers are off-net customers requiring peering of CENs belonging to different carriers. This peering of CENs is the topic of Chapter 5.

5

PEERING CARRIER ETHERNET NETWORKS

Coming together is a beginning, staying together is progress, and working together is success.

Remarks by Henry Ford

As we discussed in previous chapters, Ethernet was originally designed as a local area network (LAN) communication protocol, allowing computers and nodes to be interconnected within a small local network of an organization. We also discussed how over the years, Ethernet has evolved to become such a popular technology that it became the default data link layer (layer 2 of the OSI model) mechanism for data transport over Carrier Ethernet networks (CENs). MEF[36] estimates that enterprise demand for Ethernet services will experience a continuous and rapid growth reaching close to 50 billion USD by 2016. This growth will lead to, as we recognized in the last chapter, situations where potential subscribers for Ethernet services will have locations that are not all served by a single CEN operator. In order for such a subscriber to obtain services, multiple CEN operators will need to peer their CENs in order to support exchange of data from all the subscriber's User Network Interfaces (UNIs). Presently, this peering is being done based on TDM-based meet points which require protocol translations and supporting multiple technologies. In order to overcome these issues and leverage Ethernet technology, MEF has proposed an Ethernet-Access type of service using External Network–Network Interface (ENNI) for peering of CENs of different operators with separate administrative domains. The E-Access service type is defined in MEF 33[36] and MEF 51[36]. This service type is based on operator virtual connection (OVC). These OVCs provide connectivity between ENNI and UNIs. We covered UNIs in Chapter 4. In this chapter, we will cover E-Access service type and the associated ENNI reference model. It is important to note that the S-tag plays an important role in this peering of CENs, not only for switching decisions but for keeping traffic from different subscribers separate and also in translating class of service (CoS) regime of one CEN operator to the CoS regime of the interconnecting CEN operator. In view of this importance of the S-tag, we will start the chapter with a revisit to description of tags and bridging techniques. Following this, we will cover terminology and architecture associated with E-Access service type and then cover different services associated with E-Access service type and the attributes and parameters associated with them. Following this, the chapter will describe the management of QoS and service management in peering CENs. In the end, the chapter will provide a transition to the next chapter describing the need to exchange standardized ordering data between CEN operators to accomplish peering of CENs.

5.1 Revisiting Bridging Techniques and Tags

In Section 3.1 in Chapter 3, it was mentioned that IEEE subcommittee began its work related to LAN standards in February of 1980, and therefore all standards coming out of this subcommittee are numbered starting with 802. The standards related to LAN bridging are numbered 802.1 (please refer to Table 4.1 in Chapter 4), standards related to LLC sublayer of the data link layer (layer 2 of the OSI model) are numbered 802.2, and the standards related to media access control (MAC) sublayer of the data link layer and PHY layer numbered 802.3 (please refer to Table 3.3 in Chapter 3). We also covered in Chapters 3 and 4 all the standards related to VLANs (Q-tag), PB (Q-in-Q), Provider Backbone Bridge (PBB; MAC-in-MAC), and PBB—Traffic Engineering (PBB-TE). Due to the importance of tags, particularly S-tag in peering CENs in keeping traffic from different subscribers separate in ENNI and also in translating CoS regime of one CEN operator to the CoS regime of the interconnecting CEN operator, it is important to revisit bridging techniques and tags. Fig. 5.1 shows IEEE standards and the corresponding evolution in Ethernet frames.

As we discussed in earlier chapters, Ethernet switches (also known as bridges) are layer 2 devices connecting different LANs. Switches have a MAC layer at each port. An Ethernet switch implements MAC address learning which allows it to gradually build a forwarding database (FDB) consisting of MAC addresses of all the end stations connected on the LAN and the ports on the bridge by which they can be reached. It builds this table by examining the Ethernet frame whenever it receives it and stores the source MAC address and the port on which the frame arrived in a source address table (SAT). When the bridge receives an Ethernet frame for a destination which is in the FDB, then it will send the frame to the port from which the destination can be reached. If the destination address of an outgoing frame is not in its SAT, a switch acts

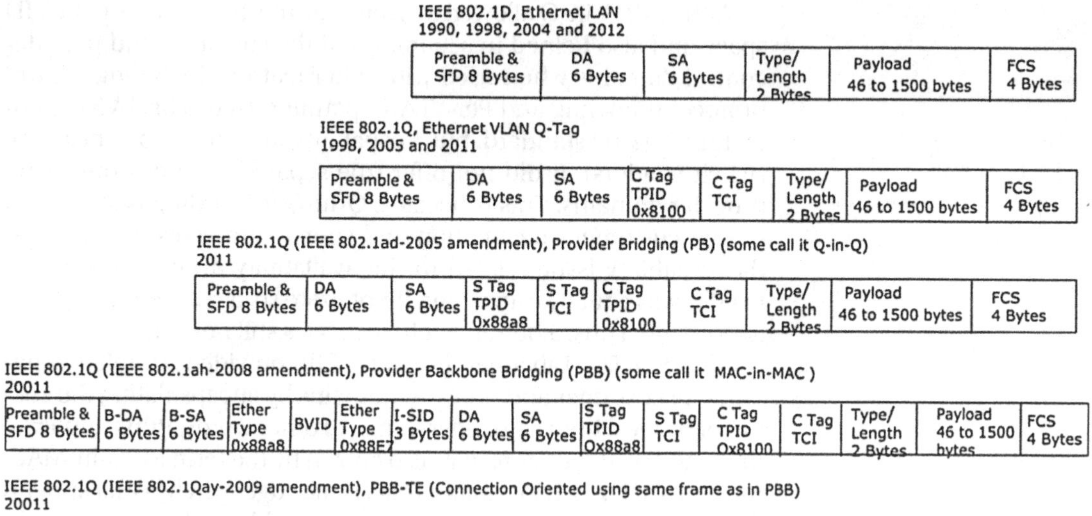

IEEE 802.1D, Ethernet LAN
1990, 1998, 2004 and 2012

Preamble & SFD 8 Bytes	DA 6 Bytes	SA 6 Bytes	Type/ Length 2 Bytes	Payload 46 to 1500 bytes	FCS 4 Bytes

IEEE 802.1Q, Ethernet VLAN Q-Tag
1998, 2005 and 2011

Preamble & SFD 8 Bytes	DA 6 Bytes	SA 6 Bytes	C Tag TPID 0x8100	C Tag TCI	Type/ Length 2 Bytes	Payload 46 to 1500 bytes	FCS 4 Bytes

IEEE 802.1Q (IEEE 802.1ad-2005 amendment), Provider Bridging (PB) (some call it Q-in-Q) 2011

Preamble & SFD 8 Bytes	DA 6 Bytes	SA 6 Bytes	S Tag TPID 0x88a8	S Tag TCI	C Tag TPID 0x8100	C Tag TCI	Type/ Length 2 Bytes	Payload 46 to 1500 bytes	FCS 4 Bytes

IEEE 802.1Q (IEEE 802.1ah-2008 amendment), Provider Backbone Bridging (PBB) (some call it MAC-in-MAC) 20011

Preamble & SFD 8 Bytes	B-DA 6 Bytes	B-SA 6 Bytes	Ether Type 0x88a8	BVID	Ether Type 0x88E7	I-SID 3 Bytes	DA 6 Bytes	SA 6 Bytes	S Tag TPID 0x88a8	S Tag TCI	C Tag TPID 0x8100	C Tag TCI	Type/ Length 2 Bytes	Payload 46 to 1500 bytes	FCS 4 Bytes

IEEE 802.1Q (IEEE 802.1Qay-2009 amendment), PBB-TE (Connection Oriented using same frame as in PBB) 20011

Figure 5.1 Evolution of Ethernet frame with bridging techniques.

like a layer 1 repeater by sending a copy of the frame to all output ports. This is known as flooding.

In 1998, the IEEE 802.1Q standard introduced the Q-tag. This is also known as C-tag. This is a new frame field that includes an explicit 12-bit VLAN identifier (VID). A VLAN is essentially a logical partition of the network. VLANs were introduced to split the LAN's broadcast domain to reduce the scope of flooding.

With the evolution of Ethernet switches, carriers started deploying them in CENs and in early deployments, VLANs were seen by carriers as the natural way to differentiate customer networks while maintaining a cheap end-to-end Ethernet infrastructure. In this setting, service providers assigned a unique 12-bit VID field within the Q-tag to each customer network. Ethernet switches added the Q-tag at the ingress node and removed it at the egress node. This use of VLANs quickly ran into scalability issues. The VID's 12 bits limited the number of supported customers to a maximum of 4094 (excluding reserved VIDs "0" and "4095"). In addition, the customers required the same VID field to partition and manage their own networks, leading to further customer–provider interoperability issues.

In an effort to mitigate scalability issues faced by the carriers, an additional Q-tag was introduced in IEEE 802.1ad amendment for Provider Bridge (PB). This resulted in two separate tags specifically meant to be used by customers (C-tag) and service providers (S-tag). This is often referred to as Q-in-Q. In IEEE 802.1ad, the CFI field in tag TCI (please refer to Fig. 3.7 in Chapter 3) is also replaced by a Drop Eligibility Indicator (DEI), thus increasing the functionality of the PCP field.

Although Q-in-Q allowed to overcome the limitations on the VID space and also helped in separation of the customer and provider control domains when used with other features including control protocol tunneling and Per-VLAN Spanning Tree (This PVST is not an IEEE 802.1Q standard, but a capability introduced by one of the switch vendors), it did not offer true separation of customer and provider domains. This is because Q-in-Q forwarding is still based on the customer's destination and source addresses. In addition, the scalability issue related to the availability of only 4094 VLAN tag IDs was equally important, if not more so, a reason that needed addressing. Thus, a better mechanism was still needed.

The IEEE 802.1ah amendment for PBB provided complete separation of customer and provider domains by encapsulating the customer frame within a provider frame. In other words, PBBs duplicate the MAC layer; therefore, it is also known by the term MAC-in-MAC. As shown in Fig. 5.1, PBBs act as backbone edge switches that append their own provider backbone-destination address (B-DA) and backbone-source address (B-SA) as well as a backbone-VID (B-VID). A new 24-bit field called the service ID (I-SID) is also introduced to identify a customer-specific service instance. This 24-bit I-SID field allows up to 16 million service instances to be defined for each of the B-VID. This eliminates any concern about running out of VID space.

The main components of the PBB header, shown in Fig. 5.1, are:
1. Header components related to backbone:
   a. B-DA (6 bytes);
   b. B-SA (6 bytes);
   c. EtherType of 0x88A8 (2 bytes); and
   d. B-tag/B-VID (2 bytes), this is the backbone VLAN indicator.
2. Header component related to Service encapsulation:
   a. EtherType of 0x88E7 (2 bytes);
   b. flags that contain priority, Drop Eligible Indicator (DEI) and No Customer Address indication (e.g., operations, administration, and management [OAM] frames); and
   c. I-SID, the service identifier (3 bytes).
3. Header component related to original customer frame:
   a. customer destination address (6 bytes);
   b. customer source address (6 bytes);
   c. EtherType 0x8100 (2 bytes);
   d. customer VID (2 bytes);
   e. EtherType (e.g. 0x0800); and
   f. customer payload.

The bridges (Ethernet switches) in the PBB domain switch based on the B-VID and B-DA values, which contain 60 bits total. Bridges learn based on the B-SA and ingress port value and hence are completely unaware of the customer MAC addresses. I-SID allows to distinguish the services within a PBB domain.

Next improvement came when PBB-TE telecommunications networking amendment was approved as IEEE 802.1Qay in 2009. PBB-TE is connection-oriented approach to make Ethernet service carrier grade. Although PBB (MAC-in-MAC) was a major step forward in complete separation of customer and provider domains, PBB forwarding is based on spanning tree performed exactly as in PB (Provider Bridging). In PBB, bridges (Ethernet switches) learn backbone addresses, flood when an unknown address is identified, and use the same xSTP-based resiliency mechanisms as in PB. Predefined backup paths are not supported, and therefore convergence time depends on the topology and xSTP message rate. PBB-TE eliminates these limitations. It is based on the layered VLAN tags and MAC-in-MAC encapsulation defined in IEEE 802.1ah for PBB, but it differs from PBB in eliminating flooding, dynamically created forwarding tables, and spanning tree protocols. Compared to PBB and its predecessors, PBB-TE behaves more predictably, and its behavior can be more easily controlled by the network operator, at the expense of requiring up-front connection configuration at each bridge along a forwarding path. A service is identified by an I-SID, and each service is associated with a PBB-TE trunk. Each PBB-TE trunk is identified by a triplet of B-SA, B-DA, and B-VID. The B-SA and B-DA identify the source and destination bridges, respectively, that are the end points of the trunk. The B-VID is a backbone-VLAN identifier that is used to distinguish different trunks to the same destination. The management system configures the PBB-TE trunks on all the edge and core bridges by creating static FDB entries. The management system is also responsible for ensuring that there are no forwarding loops. The backbone edge bridges map frames to and from an I-SID and perform the MAC header encapsulation and decapsulation functions. The core bridges act as transit nodes. The frames are forwarded based on outer B-VID and B-DA. Forwarding is based on the static FDB entries, and dynamic MAC learning is not used. Any incoming broadcast or multicast frames are either dropped or encapsulated as unicast within the trunk. All Destination Lookup Failure packets are dropped instead of being flooded. By eliminating any broadcasting or flooding, and by using only the loop-free forwarding paths configured by management, there is no longer any need to use a spanning tree protocol. Path protection is provided by configuring one working and one protection B-VID for each backbone service instance. PBB-TE OAM is usually based on IEEE 802.1ag. In case of work path failure, as indicated by loss of 802.1ag continuity check messages (CCMs), the source bridge swaps the B-VID value to redirect the traffic onto the preconfigured protection path within 50 ms. In addition, the B-tag, which is identical to an S-tag in format and TPID, has 3 bits of PCP for CoS identification and can use the DEI bit for color marking within the CEN, which is necessary for color forwarding to ENNI as we will see later.

It is important to point out that during the evolution of Ethernet switches, in the beginning, they were enterprise class Ethernet switches which were not carrier grade. Once switch vendors started selling carrier class PBs, the carriers deployed them in their networks; however, they had to deploy MPLS-based switches to improve scalability and also to support other protocols. These MPLS-based switches are still being used in the carrier networks. Therefore, there is this misconception that even now MPLS is needed in CENs. This is simply not true anymore because CEN technology has come a long way. A study[37] comparing performances of PBB-TE and T-MPLS showed a 10%–20% improvement in performance of PBB-TE over T-MPLS, and another report[38] states that PBB-TE equipment leverages economies of scale inherent in Ethernet, promising solutions that are 30–40% cheaper than T-MPLS networks with identical features and capabilities giving PBB-TE a better overall return on investment.

It is interesting to note that T-MPLS, as its name implies, is a derivative of MPLS that renounces all MPLS signaling features and, like PBB-TE, uses a centralized control plane to perform routing and traffic engineering. ITU-T was working on standardizing T-MPLS. T-MPLS was stripped of the characteristics which originally made it attractive to carriers namely, control-plane automation, signaling, and QoS and therefore has yet to prove its benefits for the transport network. T-MPLS OAM, defined in ITU-T Y.1711, is different from MPLS OAM and lacks powerful management tools that carriers typically expect. As a result, T-MPLS was abandoned by the ITU-T in favor of MPLS Transport Profile (MPLS-TP) in December 2008. MPLS-TP is a profile of MPLS developed in cooperation between ITU-T and IETF since 2008 as a connection–oriented packet-switched extension.

5.2 Ethernet Access–Related Terminology and Architecture

In Section 4.1 in Chapter 4, we discussed that MEF was formed in 2001 to define and standardize carrier grade services that leverage Ethernet technology in CENs. In Section 4.2 in Chapter 4, it was mentioned that MEF 6.2[36] defined three service types namely, (1) Point-to-Point (P2P) or Ethernet-Line or E-Line, (2)

Multipoint-to-Multipoint (MP2MP) or Ethernet-LAN or E-LAN, and (3) Rooted Multipoint or Ethernet-Tree or E-Tree, and MEF 33[36] introduced a fourth service type, namely, Ethernet-Access or E-Access. First three service types were covered in detail in Chapter 4. The fourth service type namely, E-Access is used by leveraging ENNI for peering CENs, and these are the topics of this chapter. The E-Access service type enables offering of CE services for subscriber locations that are not all served by a single CEN Operator. In order for such a subscriber to obtain services, multiple CEN operators will need to peer their CENs to support exchange of data from all the subscriber's UNIs. This peering of CENs of different operators requires ENNI and leverages E-Access service type. It is important to note that the descriptions in Chapter 4 of E-Line, E-LAN, E-Tree, UNI, Ethernet virtual connection (EVC), QoS, CoS, bandwidth profile, traffic engineering, performance attributes, and service management are still applicable in this chapter also. However, E-Access introduces additional specifications, and these in turn introduce new terms and definitions. Table 5.1 shows some of the MEF specifications that are relevant to E-Access service type for peering CENs. It should be noted that these specifications are additionally relevant to those listed in Table 4.2 in Chapter 4. Table 5.1 shows that MEF 26.1 defined ENNI and OVC service attributes and parameters which have been revised in MEF 26.2. MEF 33

Table 5.1 Sample of MEF Standards for E-Access Service Type

MEF Specification	Year	Objectives	Notes
6.1	2008	Ethernet service definitions	Superseded by MEF 6.2 in 2014
32	2011	Service protection across external interfaces	
23.1	2012	CoS mapping between CENs	
26.1	2012	External Network–Network Interface (ENNI), ENNI and OVC attributes and parameters	
33	2012	Ethernet Access services definitions, attributes, and parameters. Defined Access E-line service type	
34	2012	Abstract test suites for Ethernet Access services	
6.2	2014	Ethernet EVC service definitions	

Continued

Table 5.1 Sample of MEF Standards for E-Access Service Type—continued

MEF Specification	Year	Objectives	Notes
51	2015	OVC service definitions and defined new Access E-LAN and E-Transit service types	OVC service attributes defined in this document that go beyond MEF 26.1. This specification also defines E-Transit service type.
23.2	2016	Defined new PT 0.3 and CPOs for multipoint services	Supersedes MEF 23.1
26.2	2016	Revised bandwidth profile attributes and removed mandatory color awareness requirement at ingress ENNI	Supersedes MEF 26.1

introduced Access Ethernet services and defined Access E-Line service type based on attributes and parameters defined in MEF 26.1 and MEF 26.2.

In addition, MEF 51 introduced Access E-LAN and E-Transit service types based on OVC-type attribute at the external interface. The E-Transit service type is for OVC connecting two ENNIs. The MEF 26.1 and MEF 33 specifications have introduced new terms like Ethernet-Access service type, Ethernet Access Private Line (Access-EPL), Access-Ethernet Virtual Private Line (EVPL), ENNI, OVC, access provider, subscriber's point of view, and service provider's point of view. It is important to understand these terms before going in to details of Ethernet Access service type and services defined by it and their attributes and parameters. Fig. 5.2 shows the topology of a P2P service provided by a service provider to a customer in order to connect customer's site A with another geographically distributed site Z, so that the end stations in these two locations are made part of the same LAN thus enabling the end stations to exchange data. The figure shows that in order to connect to site Z located in another operator's territory, peering of CENs based on ENNI is needed.

End stations in subscriber's network in site A are connected to port 1 on the customer's Ethernet switch known as customer edge (CE 1) on the access link. Customer configures the switch to direct all frames from port 1 to port 4 which is then connected

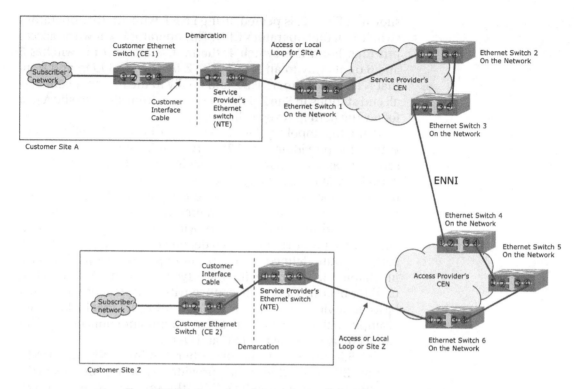

Figure 5.2 Topology of a point-to-point service requiring peering CENs.

to the port 1 on another Ethernet switch known as network termination equipment (NTE), also known as network interface device, provided by the service provider[1] and located at the customer premises. Port 1 on this NTE is the demarcation point between customer and the service provider. This NTE in general has many ports but in this case as shown in Fig. 5.2, only one port is provided to the customer. NTE then is configured by the service provider to direct all frames from port 1 to port 4. This port is then connected by a local loop to an Ethernet network switch in the CO of the service provider. Common practice is to connect NTE to an Ethernet multiplexer, so that traffic from various customers in that location or building can be multiplexed on the one local loop and sent to CO. The frames from port 1 on the network switch 1 are directed to port 4 which is connected to another Ethernet switches 2 and finally switched to Ethernet switch 3 on the service provider's CEN. This Ethernet switch 3, as

---

[1] In this chapter, we will assume that service provider is one of the operators. This, however, may not be the case always. A service provider, in general, is an entity that commissions service from operators.

shown in Fig. 5.2, is peered, using ENNI, with another operator's switch 4 on that operator's CEN. This operator is known as access provider. From this switch 4, the frames are sent to switches 5 and 6 on its way to an NTE at site Z on the local loop and from that NTE to customer's CE 2 and finally to the hub that connects all end stations in Site Z. Ethernet frames from site Z to site A just follow the path in reverse direction.

Using this topology, Fig. 5.3 shows various MEF-defined constructs for providing this P2P service utilizing Ethernet-Access service type. These constructs include UNI 1, OVC 1, ENNI, OVC 2, UNI 2, and EVC. In Chapter 4, we had described that MEF uses interface type and connectivity to categorize Ethernet services. UNI is one interface type and represents a service demarcation between end user and service provider. UNIs are billable items and have many attributes which depend on the service types with each attribute defined by one or more parameters as described in Chapter 4. The other interface type is called ENNI which is used for peering CENs. The connectivity requirement is the other basis in service categorization used by MEF and that includes, for example, P2P or MP2MP or rooted-multipoint connectivity. This connectivity is defined by EVC and OVC.

Fig. 5.3 shows that the subscriber requires UNI 1 and UNI 2 and an EVC from the service provider. However, because UNI 2 is not in service provider's CEN, the service provider in turn first buys an E-Access service from an access provider which predicates establishing an ENNI that peers service provider's CEN with the CEN of the access provider, and then the access provider provides UNI 2 to service provider. The service provider provisions UNI 1 it its territory. After this, service provider provides connectivity between UNI 1 and ENNI with OVC

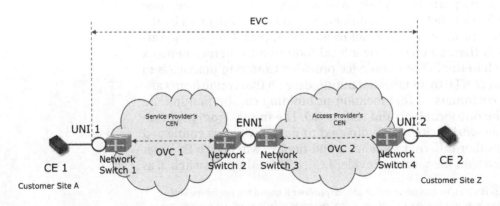

Figure 5.3 Logical view of the peering CENs using Ethernet-Access service type.

1 and similarly access provider connects UNI 2 to ENNI with OVC 2. Combined OVC 1 and OVC 2 in turn results in the EVC for the subscriber over the peered CENs. The subscriber only deals with service provider and service provider in turn buys access from access provider on behalf of the subscriber. In this model, the subscriber does not deal with access provider at all and does not even know that the service is provided over peered CEN using ENNI.

Although UNI and ENNI are both interfaces and represent Ethernet ports that demarcate buyers and sellers of services, there are some very basic differences vis-à-vis capabilities, and deployments are concerned. These differences are shown in Table 5.2.

Table 5.2 Comparison of UNI and ENNI Interfaces

Capability	UNI	ENNI
Demarcation	Between subscriber and operator	Between service providing operator and access providing operator
Connection	Connects subscriber CE to operator's edge bridge (switch) on CEN	Connects service providing operator's CEN edge/core switch to access providing operator's CEN edge/core bridge (switch)
Ingress Ethernet frame format supported	Untagged or single C-tagged as per IEEE 802.1Q	Double tagged as per IEEE 802.1ad (Q-in-Q)
Virtual connectivity type	EVC or OVC	OVC
Number of supported service instances	Usually small numbers of EVCs or OVCs	Varies between 1 and 1000s OVCs
Ethernet bandwidth	Typically 10/100/1000 Mbps	1, 10, 100 Gbps (even higher in future)
Multiplexing of virtual connections	May or may not	Always
Ethernet frame size	64–1522 bytes	1526 bytes (minimum)
CoS identifier	IEEE 802.1Q C-tag PCP (p bit) or DSCP	IEEE 802.1ad S-tag PCP or DSCP

OVC, operator virtual connection; *UNI*, User Network Interface.

The major difference between EVC and OVC is that EVC is between UNIs and OVC is between any set of external interfaces (UNI and ENNI), but at least one has to be ENNI. EVCs are important from subscriber's point of view, and OVCs are important from operator's point of view. This is shown in Fig. 5.4.

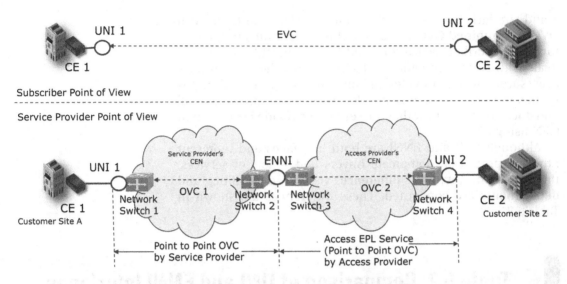

Figure 5.4 Subscriber and service provider points of views.

Fig. 5.4 shows that a subscriber buys UNI 1 and UNI 2 and an EVC from a service provider. Since UNI 2 is not in the service provider's territory, the service provider turns to access provider and establishes an ENNI and then buys UNI 2 and then access provider connects UNI 2 to ENNI with a virtual connection shown in Fig. 5.4 as OVC 2. On its part, the service providing operator provides to the subscriber UNI 1 and connects it to the ENNI with a virtual connection OVC 1. These two OVCs namely, OVC 1 and OVC 2, together result in the subscriber's EVC. As shown in Fig. 5.4, the subscriber's point of view consists of UNI 1, UNI 2, and EVC because that is what the subscriber is buying from the service provider. The service providing operator's point of view consists of UNI 1, UNI 2, ENNI, OVC 1, OVC 2, and EVC because the service provider is responsible for all these. The access providing operator's point of view, on the other hand, is limited to UNI 2, OVC 2, and ENNI. It should be noted that MEF 33 has defined ENNI as a reference point representing the boundary between two operator CENs that are operated as separate administrative domains. The two parties to an ENNI are "network operators" who may or may not be the Service Provider or the Access Provider, however as stated before, in this book, we will assume that the two parties are service provider and access provider. MEF 51[36] has defined a new E-Transit service type. E-Transit service type for any OVC service associates only ENNIs. MEF has left the actual implementation of ENNI to service provider and access provider. We will discuss, as part of discussions on next steps, in Chapter 9, some of the operational issues encountered in the implementation of ENNI that needs to be addressed or standardized or coordinated between operators.

When redundancy is needed at ENNI, then two physical links are provided at ENNI and that requires Ethernet switches on both service provider side and access provider side to support link aggregation within one Link Aggregation Group (LAG) across ENNI with one active link and another link in standby mode. When LAG is deployed at ENNI then Link Aggregation Control Protocol (LACP) must be used to monitor the status of active and standby links and to failover from active to standby link in case of issues with the active link.

Now that we understand terms like Ethernet-Access service type, ENNI, OVC, service provider, access provider, subscriber's point of view, and service provider's point of view, next important item to understand is the role of S-tag in Ethernet frame in passing the data traffic between service providing operator and access providing operator. This is extremely important because data traffic of many customers travels on ENNI as shown in Fig. 5.5, and they have to be prevented from intermingling.

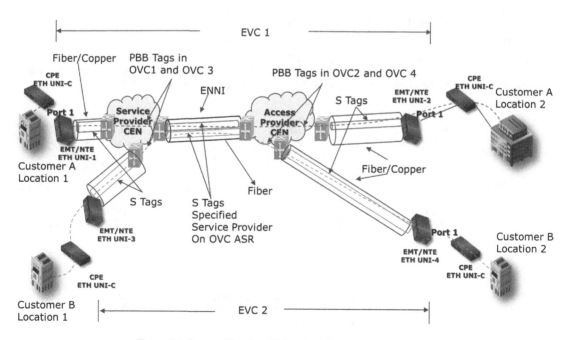

Figure 5.5 Data traffic of multiple subscribers traversing ENNI.

In addition, S-tag in Ethernet frame is used to coordinate CoS and drop eligibility (DEI) between service providing and access providing operators in order to deliver the agreed-upon SLAs to the end customer. This coordination of CoS is required because each operator may offer different CoS, and at ENNI, they have to be mapped correctly. This is covered in more detail later when we discuss QoS treatment on peering CENs. This importance of S-tag in ENNI for peering CENs is the main reason we revisited bridging techniques and tags in Section 5.1. Now, the combined learning from Figs. 5.1 and 5.4 is shown in Fig. 5.6 to help explain the role of S-tag in ENNI. For discussion, let us assume that Ethernet frame starts at Site A from CE 1 going to site Z and on to CE 2. In this example of P2P connection, the customer may have C-tag as per IEEE 802.1Q or may not have C-tags. At NTE, the service provider encapsulates the Ethernet frame with an S-tag as per IEEE 802.1Q (with merged IEEE 802.1ad amendment for PB). This is to prevent intermingling of Ethernet frames of different customers or of different virtual circuits. At the edge switch/bridge on the service provider's CEN, the S-tag is popped and the Ethernet frame is encapsulated in PBB-based MAC-in-MAC frame as per IEEE 802ah shown in Fig. 5.6. Also, in the CEN, the service provider implements

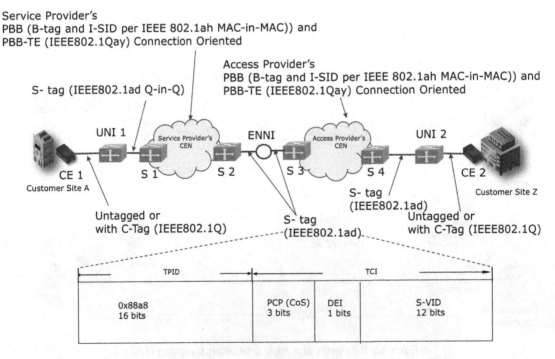

Figure 5.6 Use of bridging and tag in Ethernet-Access service type using ENNI.

connection-oriented path as per PBB-TE based on IEEE 802.1Qay. When the Ethernet frame reaches switch S2 shown in Fig. 5.6, the PBB frame is popped. This is because PBB and PBB-TE are limited to service provider's CEN and cannot be carried forward to access provider's CEN. As the Ethernet frame egresses S2 switch, a new S-tag is inserted in the Ethernet frame. This S-tag prevents intermingling of Ethernet frames of various virtual connections traversing the ENNI. As the Ethernet frame reaches S3 switch on the access provider's CEN, the S-tag is popped and the Ethernet frame is encapsulated in PBB-based MAC-in-MAC frame and switched on access provider's CEN as per PBB-TE connection–oriented path. When the Ethernet frame reaches the edge switch S4, the PBB frame is popped and an S-tag is pushed in. Next, when the Ethernet frame reaches UNI 2, S-tag is popped and the customer's Ethernet frame with or without C-tags is sent to customer's edge device CE 2. Ethernet frames from CE 2 to CE 1 will follow the path in reverse direction.

With this information, on bridging techniques and tags and definitions of Ethernet-Access service type, ENNI, OVC, service provider, access provider, subscriber's point of view, service provider's point of view, and access provider's point of view, application of S-tag at ENNI and the purpose it serves, we can now examine the attributes and parameters associated with ENNI and OVCs for the Ethernet-Access service types. It is important to recall the descriptions especially of Ethernet services framework from Section 4.2 in Chapter 4 because those descriptions are also very helpful at this point. It is also important to recall from Chapter 4 that specific services can be defined by assigning different values to parameters related to each attribute.

5.3 Attributes and Parameters Related to Ethernet Access Services

As discussed in last section, subscriber's point of view involves UNIs and EVCs. Their attributes and parameters have already been described in Chapter 4. From operator's point of view, there are OVCs, ENNI, and OVC end points at UNI as well as ENNI as shown in Fig. 5.4 and simplified further in Fig. 5.7. Here, OVC 1 has two OVC End Points (O EPs) 1 and 2, OVC 2 also has two end points O EP 3 and 4, and ENNI has two end points O EP 2 and 3. Based on Fig. 5.7, it is clear that attributes and parameters need to be defined not only at ENNI and OVC but also at OVC end points both at UNI and ENNI, so that OVCs can be properly concatenated to get the proper EVC that the

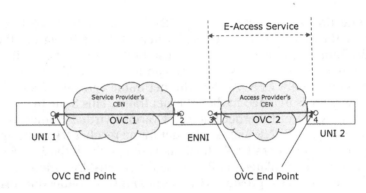

Figure 5.7 ENNI, OVC, and OVC end points.

subscriber has ordered as well as to ensure SLAs. MEF 26.1 has in fact defined attributes and parameters for ENNI, OVC, O EPs per ENNI, and O EPs per UNI. MEF 51[36] has defined E-Transit service type to interconnect operator CENs to the left and right of the ENNI shown in Fig. 5.7. Since ENNI is an interface between CENs of two operators, ENNI will have two sets of attributes: one for service provider and the other for access provider. Parameters for these ENNI attributes may or may not have identical values. At this point, it is important to understand two additional terms namely, ingress ENNI frame and egress ENNI frame. With reference to Fig. 5.7, when an Ethernet frame is traveling from O EP 2 to O EP 3, it is called egress ENNI frame from the perspective of transmitting operator. This same Ethernet frame coming to O EP 3 is called ingress ENNI frame from the perspective of the receiving operator.

The Ethernet service attributes and their associated parameters based on MEF 26.1[36] for ENNI are shown in Table 5.3.

Table 5.3 ENNI Service Attributes

Attribute	Type of Parameter Value
Operator ENNI ID	A String that is unique across operator CEN
Physical layer	A standard Ethernet PHY per IEEE 802.3–2012, for example, 1000BASE-SX, 1000BASE-LX, 1000BASE-T, 10GBASE-SR, 10GBASE-SR, 10GBASE-LX4, and so forth. This specifies both the type of physical interface and speed. The physical Layer MUST operate in full-duplex mode.
Frame format	Must have S-tag as per IEEE 802.1ad.

Table 5.3 ENNI Service Attributes—continued

Attribute	Type of Parameter Value
Number of links	≥1
Protection mechanism	None or link aggregation
ENNI max transmission unit size	≥1526
End point map	A Table showing VID value, end point identifier and end point type (normally OVC)
Maximum number of OVCs	Integer greater than or equal to 1
Maximum number of OVC end points per OVC	Integer greater than or equal to 1

Reproduced with permission of the Metro Ethernet Forum.

The Ethernet service attributes and their associated parameters based on MEF 26.1[36] for OVC are shown in Table 5.4.

Table 5.4 OVC Service Attributes

Attribute	Type of Parameter Value
OVC ID	A string that is unique across operator CEN
OVC type	Point-to-point, multipoint-to-multipoint, or rooted multipoint. A non-null string not greater than 45 characters
OVC end point list	A list of OVC end point ID, OVC end point role
Maximum number of UNI OVC end points	Integer greater than or equal to 0
Maximum number of ENNI OVC end points	Integer greater than or equal to 1. If the sum of UNI OVC end points and ENNI OVC end points is less than 2, then the OVC cannot carry traffic across CENs of two different operators.
OVC maximum service frame size	≥1526
CE-VID preservation	Enabled (Yes) or disabled (No)
CE-VLAN CoS preservation	Enabled or disabled
S-VID preservation	Enabled or disabled
S-VLAN CoS preservation	Enabled or disabled
Color forwarding	Yes or No
Service-level specification	Frame delay, frame delay variation, frame loss, availability specified here (refer to Section 7.2.16 in Chapter 7 of MEF 26.1 for additional details)
Unicast data service frame delivery	Deliver unconditionally or deliver conditionally
Multicast data service frame delivery	Deliver unconditionally or deliver conditionally
Broadcast data service frame delivery	Deliver unconditionally or deliver conditionally

Continued

Table 5.4 OVC Service Attributes—continued

Attribute	Type of Parameter Value
Layer 2 control protocols processing (L2CP) that are tunneled by OVC	Process, discard, or pass the following L2CP service frames per MEF 6.1.1 and MEF 10.2 (refer Section 7.2.20 in Chapter 7 of MEF 26.1 for details): 1. IEEE 802.3x MAC control 2. Link aggregation control protocol (LACP) 3. IEEE 802.1x port authentication 4. Generic attribute registration protocol (GARP) 5. STP 6. Protocols multicast to all bridges in a bridged LAN

Reproduced with permission of the Metro Ethernet Forum.

The Ethernet service attributes and their associated parameters for O EPs per ENNI based on MEF 26.1[36] are shown in Table 5.5.

Table 5.5 OVC End Point Per ENNI Service Attributes

Attribute	Type of Parameter Value
OVC end point ID	A string that is unique across operator CEN
Trunk identifier	N/A for root or leaf OVC. Has a value trunk OVC end point only (refer Section 7.3.2 in Chapter 7 of MEF 26.1).
Class of service (CoS) identifiers	CoS identified by the S-tag's PCP value in the frame.
Ingress bandwidth profile per OVC end point	No, if no policing of ingress ENNI frame is needed or specify CIR, CBS, EIR, EBS, CM, CF as per Section 7.6.1 in Chapter 7 of MEF 26.1 if policing of ingress ENNI frame is needed
Ingress bandwidth profile per ENNI CoS ID	No or parameters as defined in Section 7.6.1 in Chapter 7 of MEF 26.1
Egress bandwidth profile per OVC end point	No, if no policing of egress ENNI frame is needed or specify CIR, CBS, EIR, EBS, CM, CF as per Section 7.6.1 in Chapter 7 of MEF 26.1 if policing of egress ENNI frame is needed
Egress bandwidth profile per ENNI CoS ID	No or parameters as defined in Section 7.6.1 in Chapter 7 of MEF 26.1

Reproduced with permission of the Metro Ethernet Forum.

The Ethernet service attributes and their associated parameters for OVC per UNI based on MEF 26.1[36] are shown in Table 5.6. It should be noted that an OVC can associate only O EPs at a UNI, and so these attributes can also be referred to as O EPs per UNI service attributes.

By applying different parameter values to the attributes for ENNI, OVC, O EP per ENNI, and OVC per UNI given in Tables 5.3–5.6, different services of Ethernet-Access service type can be constructed for peering CENs using ENNI.

Table 5.6 OVC Per UNI Service Attributes

Attribute	Type of Parameter Value
UNI OVC end point ID	A string formed by concatenation of UNI ID and the OVC ID
OVC end point map	A list of one or more CE-VID values
Class of service (CoS) identifiers	CoS identified by the C-tag's PCP value in the frame (refer Section 7.5.3 in Chapter 7 of MEF 26.1 for additional details).
Ingress bandwidth profile per OVC end point at a UNI	No, if no policing of ingress frame is needed or specify CIR, CBS, EIR, EBS, CM, CF as per Section 7.5.5 in Chapter 7 of MEF 26.1 if policing of ingress frame is needed
Ingress bandwidth profile per UNI CoS ID	No or parameters as defined in Section 7.5.5 in Chapter 7 of MEF 26.1
Egress bandwidth profile per OVC end point at a UNI	No, if no policing of egress frame is needed or specify CIR, CBS, EIR, EBS, CM, CF as per Section 7.5.6 in Chapter 7 of MEF 26.1 if policing of egress frame is needed
Egress bandwidth profile per UNI CoS ID	No or parameters as defined in Section 7.5.6 in Chapter 7 of MEF 26.1

Reproduced with permission of the Metro Ethernet Forum.

This background information of the Ethernet services framework for Ethernet-Access service type and the associated attributes and parameters will help in understanding Access-EPL and Access-EVPL services which are described in the next section. These services are derived from Ethernet-Access service type.

5.4 Ethernet Access Services on Peering Carrier Ethernet Networks

We know from Chapter 4 that subscriber Ethernet services (E-Line, E-LAN, E-Tree) are either port based or VLAN based. In port-based or "Private" service, the all-to-one bundling attribute, for UNIs in the service, is ENABLED. In case of VLAN-based or "Virtual Private" service, the all-to-one bundling attribute, for UNIs in the service, is DISABLED. Virtual Private services can be service multiplexed at a UNI (i.e., more than one service can terminate at the UNI) since the customer VLAN is used to map frames to services. Adopting that definition for E-Access services requires carefully differentiating service provider UNI from access provider UNI. An E-Access service is used to extend a subscriber service to a remote access providing operator UNI, and that UNI must have the same value for all-to-one bundling as the other service providing operator's UNIs in the service. This results in two different E-Access services: the Access-EPL which is used to extend a private subscriber service and Access-EVPL which is used to extend a virtual private subscriber service. MEF 33[36] defines these two Ethernet services for the E-Access Service type slightly differently based on the access provider UNI. According to MEF 33[36], the services where service frames at the (access provider) UNI can be mapped to only a single OVC end point are referred to as "Private" or port-based services or Access-Ethernet Private Line or Access-EPL service. The services where frames are mapped to one member of a set of OVC end points or to one member of a set of OVC end points and EVCs are referred to as "Virtual Private" or VLAN-based services or Access-Ethernet Virtual Private Line or Access-EVPL service. In the agreement between the service provider and access provider, these services are primarily distinguished by three attributes. The first one is the operator UNI-related attribute called "Maximum number of OVCs per UNI." This must be 1 for the Access-EPL and maybe be ≥1 for Access-EVPL. This reflects the fact that private services cannot terminate on service multiplexed UNIs but virtual private services may or may not. The second attribute is the OVC per UNI-related attribute called "O EPs Map" which must map all values from 1 to 4095 for Access-EPL reflecting the all-to-one bundling characteristic of the subscriber service. This attribute can have any non-null proper subset of CE-VLAN values for Access-EVPL. The third attribute is the operator UNI-related attribute called "CE-VID for untagged and priority-tagged frames." This attribute must contain a value from 1 to 4095 for Access-EPL (it does not matter what value since all CE-VIDs are mapped to the same service) but for Access-EVPL, this attribute only contains a value if untagged and priority-tagged frames are to be carried by the service.

It is important to reiterate that Access-EPL and Access-EVPL services are within access providing operator's CEN and are needed to provide service to off-net customers using ENNI-based peering between CENs of service providing and access providing operators. These services do not concern with the end subscriber or customer as we discussed in prior sections in this chapter. According to MEF 33, Access-EPL service is needed to provide subscribers EPL and Ethernet Private-LAN (EP-LAN) services over peering CENs, and Access-EVPL service is needed for providing EVPL and EVP-LAN services to subscribers over peering CEN services.

5.4.1 Access-Ethernet Private Line Service

An Access-EPL service must use a P2P OVC that associates one O EP at a UNI and one O EP at an ENNI as shown in Fig. 5.8. The Ethernet frame's header and payload upon ingress at the UNI 2 is delivered unchanged to the ENNI, with the addition of an S-tag at O EP 3. The frame's header and payload upon ingress at the ENNI is delivered unchanged to the UNI except for the removal of the S-tag at O EP 2. In the reverse direction, the S-tag is added at O EP 2 and removed at O EP 3. The FCS of the Ethernet frame is recalculated when an S-tag is added or removed. Also, ENNI has to accept Ethernet frames that are 1526 bytes or larger. A Service Provider can buy the Access-EPL service from an access provider to deliver the port-based Ethernet services, including Ethernet Private Line (EPL), and EP-LAN defined in MEF 6.2[36], to a subscriber. Ethernet Private Tree (EP-Tree) services are not yet supported by Access-EPL in MEF

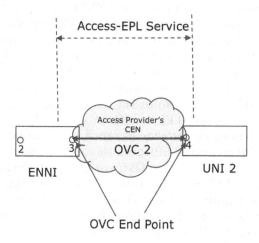

Figure 5.8 Access-EPL service.

33. Since all service frames at the UNI are mapped to a single O EP in this Access-EPL service, there is no need for coordination between the subscriber and service provider on a detailed C-tag VID/EVC map for each UNI. However, the service provider and access provider need to coordinate the value of the S-tag and other service attributes at the ENNI. The customer edge (CE) device is expected to shape traffic to the ingress bandwidth profile of the service such that all of its traffic, including certain L2CPs that require delivery for proper operation, is accepted by the service, else the policing at ingress to UNI may drop frames. This is because all service frames at the UNI are mapped to a single O EP. MEF 33 gives details of service attributes and associated parameters for Access-EPL service for UNI, OVC, ENNI, OVC per UNI, and O EPs per ENNI. Although only one UNI is shown in access provider's CEN in Fig. 5.8, there can be multiple UNIs connected to different OVCs all traversing same ENNI but using different S-tags. This case of multiple UNIs is illustrated in more detail in the section on Access-EVPL and in Fig. 5.9. In Access-EPL service, Ethernet frame's header and payload upon ingress at the UNI is delivered unchanged to the ENNI, with the addition of an S-tag. The frame's header and payload upon ingress at the ENNI is delivered unchanged to the UNI except for the removal of the S-tag. These actions require that the FCS for the frame is recalculated when an S-tag is inserted or removed.

5.4.2 Access-Ethernet Virtual Private Line Service

MEF 33 defines an Access-EVPL service that uses a P2P OVC that associates a UNI O EP and an ENNI O EP. Fig. 5.9 shows an example of the logical view of the Access-EVPL service.

An Access-EVPL can be used to create services similar to the Access-EPL; however, there are some important differences. First, with Access-EVPL, a UNI on the access provider's side can support multiple service instances, including a mix of Access and EVC services as shown in Fig. 5.9. Such configurations are not possible if Access-EPL is offered at the UNI. Second, an Access-EVPL need not provide as much transparency of service frames as with an Access-EPL because the O EP map determines which C-tags are mapped to OVCs or dropped. Because multiple instances of EVCs and Access-EVPLs are permitted, not all ingress service frames at the UNI need be sent to the same destination. In this service also Ethernet frame's header and payload upon ingress at the UNI is delivered unchanged to the ENNI, with the addition of an S-tag. In the reverse path, the frame's header

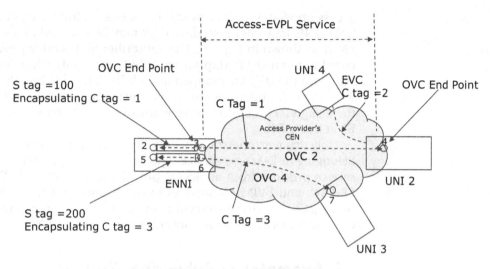

Figure 5.9 Access-EVPL service.

and payload upon ingress at the ENNI is delivered unchanged to the UNI except for the removal of the S-tag. These actions require that the FCS for the frame is recalculated when an S-tag is inserted or removed.

With Access-EVPL, the UNI can support multiple service instances, as shown for UNI 2 in Fig. 5.9, where C-tag ID of 2 is mapped to the EVC that associates UNI 2 and UNI 4. C-tag ID 1 is mapped to OVC 2 End Point 4, and associated via the Access-EVPL with OVC 2 End Point 3 at the ENNI. UNI 3 illustrates another instance of Access-EVPL connecting it to the ENNI. At ENNI, the OVC 2 End Point 3 adds S-tag ID 100 and encapsulates C-tag ID 1. On delivering the frame to O EP 2, the S-tag is popped. Similarly, for OVC 4, S-tag ID 200 is added at OVC4 End Point 6 encapsulating C-tag ID 3. On delivering the frame on ENNI to O EP 5, S-tag is popped. O EPs 3, 6 and 2, 5 represent multiplexing of OVCs at the ingress and egress ports at ENNI. A Service Provider can use the Access-EVPL service defined by MEF 33 from an access provider to deliver the two VLAN-based Ethernet services defined in MEF 6.2 namely, EVPL and Ethernet Virtual Private LAN (EVP-LAN) and supported by the ENNI defined in MEF 26.1[36]. EP-Tree services are not yet supported by Access-EVPL in MEF 33. The customer edge (CE) device is expected to shape traffic to the ingress bandwidth profile to minimize frame loss by the service. This is because frames from multiple service from UNIs in service provider's CEN may be mapped to a single ENNI and to a single O EP in Access-EVPL service in the access provider's CEN or in another scenario, frames from multiple UNIs in access provider's CEN may be mapped to a single ENNI as shown in Fig. 5.9. The subscriber and service provider coordinate an O EP Map for each OVC to specify what service frames at the UNI are mapped to each OVC. In addition, the service provider and access provider must coordinate the value of the S-tag VID and other service attributes that maps to each O EP at the ENNI.

Now that we understand Access-EPL and Access-EVPL services belonging to E-Access service type which can be purchased by a service provider from an access provider to provide EPL, EVPL, EP-LAN, and EVP-LAN services to a subscriber on ENNI-based peering CENs, in the next section, we will examine some examples of the services delivered to subscribers on peering CENs.

5.5 Examples of Subscriber Services Delivered on Peering Carrier Ethernet Networks

5.5.1 Ethernet Private Line Service on Peered Carrier Ethernet Networks

Fig. 5.10 shows subscriber and service provider's point of view (POV) for an instance of EPL service offered by the service provider using Access-EPL Service in the access provider network for peering CENs. The subscriber orders UNI 1 and UNI 2 and an EVC connecting these two UNIs from a service provider as shown in subscriber's POV.

Since UNI 2 is not in the service provider's CEN, the service provider turns to an access provider and places an Access-EPL service order to peer the CENs of service provider and the access provider. The access provider provides UNI 2 and connects it to the ENNI using P2P OVC 2 with O EP 4 on UNI 2 and O EP 3 on ENNI. The service provider in turn connects UNI 1 to ENNI using P2P OVC 1 with O EP 1 on UNI 1 and O EP 2 on ENNI. These are shown in service provider's POV in Fig. 5.10. Concatenation of OVC 1 and OVC 2 results in the EVC that subscriber ordered. Table 5.7 shows the attributes and their associated parameters that must be specified as part of the end-to-end service. Table 5.7 is organized by values that belong to the end-to-end service, that is, from subscriber's POV, values that belong to the access provider, and the values that belong to the service provider. This table is based on attributes and parameters that we

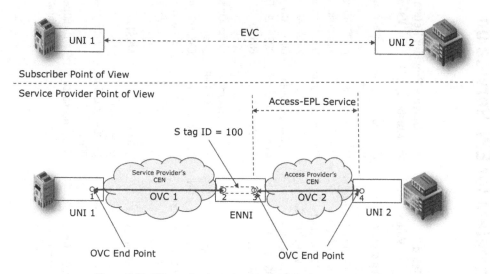

Figure 5.10 EPL service based on Access-EPL for peering CENs.

covered in Tables 5.3–5.6 and also in MEF 33. It is also important to note that MEF 6.2 does not allow bandwidth profile per EVC anymore; it only allows specifying bandwidth profile per CoS identifier.

In Table 5.7, row 1 shows how the EVC and OVC types all line up as point to point. Row 2 shows that the all-to-one bundling EVC attribute is "Yes." Row 3 illustrates the CE-VID to EVC map and the OVC End Point Maps with all CE-VLANs mapped to one OVC or EVC. Row 4 illustrates that the S-VLAN ID value for the OVC must agree for both networks in the ENNI end point maps. Row 5 shows how the ingress bandwidth profile for the EVC is realized in the two OVCs UNI end points. Here at UNI, the color mode (CM) is marked color blind because the frames arriving at UNI will not have a color assigned to them as subscriber almost always never mark frames with color (there are some situations where they might but these cases are rare), so at UNI, MEF specifications want the policer to be color blind, that is, to make no assumptions about the color of the frame. Row 6 shows that the ingress bandwidth profiles at the ENNI should be reflecting a slight CIR increase due to the per-frame S-tag overhead and also that the CM should be marked color aware because once the frames go through a policer usually at UNI, it will have a color assigned to it, and therefore the MEF 26.1 and MEF 33 specifications want all subsequent policers, which usually mean at ENNI, to take that color in to account and not,

Table 5.7 Example of Attributes and Parameters for EPL Service on Peered CENs

Row	End-to-End EPL Service Related		Access-EPL Related		Service Provider Related	
	Attribute	Value	Attribute	Value	Attribute	Value
1	EVC type	Point to point	OVC type	Point to point	OVC type	Point to point
2	All–to-one bundling	Yes	N/A		N/A	
3	CE-VID to EVC map	All service frames map to one EVC	OVC 2 end point map at UNI 2	All CE-VID values map to single OVC 2	OVC end point map at UNI 1	All CE-VID values map to single OVC 1
4	N/A		ENNI end point map	OVC 2 end point 3=SVLAN ID 100	ENNI end point map	OVC 1 end point 2=SVLAN ID 100
5	Ingress bandwidth profile per CoS identifier.	CIR=100Mbps, CBS=12K, EIR, EBS=0; CF=0	Ingress bandwidth profile per OVC 2 EP 4 at UNI 2	CIR=100Mbps, CBS=12K, EIR, EBS=0; CM=blind; CF=0	Ingress bandwidth profile per OVC 1 EP 1 at UNI 1	CIR=100Mb/s, CBS=12K, EIR, EBS=0; CM=blind; CF=0
6	N/A		Ingress bandwidth profile per OVC 2 EP 3 at ENNI	CIR should reflect S-tag overhead, CIR, CBS, EIR, and EBS same as at UNI 2; CM=color aware; CF=0	Ingress bandwidth profile per OVC 1 EP 2 at ENNI	CIR should reflect S-tag overhead, CIR, CBS, EIR, and EBS same as at UNI 1; CM=color aware; CF=0
7	CE-VID, CoS preservation	MUST be Yes	CE-VID, CoS preservation	MUST be Yes	CE-VID, CoS preservation	MUST be Yes
8	CoS ID	Based on EVC, value=M	CoS ID based S-tag PCP	All PCP values=M	CoS ID–based S-tag PCP	All PCP values=M
9	EVC performance	MFD=80ms for label=M, PT=continental	OVC performance	MFD=10ms; PT=regional	OVC performance	MFD=50ms; PT=continental

for example, promote a yellow frame to green frame in the middle. This requirement made sense since frames are normally policed and color marked at the UNI and it is desirable for the ENNI policer to take that marking into account. However, there are situations where this can cause problems such as transit services (between ENNIs) that do not support "yellow" frames. To allow for these cases, MEF 26.2 has removed the mandatory requirement for color-aware policing at the ENNI. Row 7 shows that CE-VLAN and CoS ID value preservation apply end to end. Row 8 states that the CoS ID for this case for the EVC has a CoS Label of M level, and in the subtending OVCs, this is accomplished by specifying that all PCP values are mapped to the M CoS. Row 9 illustrates that the EVC mean frame delay (MFD) performance for CoS label = M, and the performance tier (PT) of continental requires MFD of 80 ms according to the MEF 23.1. The individual OVCs have adequate performance to meet the end-to-end goal.

5.5.2 Ethernet Private-LAN Service on Peered Carrier Ethernet Networks

Fig. 5.11 shows subscriber and service provider's POV for an instance of EP-LAN service offered by the service provider using Access-EPL service for peering CENs in the access provider network. The subscriber orders UNI 1, UNI 2, and UNI 3 and an MP2MP EVC connecting these three UNIs from a service provider as shown in subscriber's POV. Since UNI 2 is not in the service provider's CEN, the service provider turns to an access provider and places an Access-EPL service order to peer the CENs of service provider and the access provider. The access provider provides UNI 2 and connects it to the ENNI using P2P OVC 2 with O EP 4 on UNI 2 and O EP 3 on ENNI. The service provider in turn connects UNI 1, UNI 3, and ENNI using MP2MP OVC 1 with O EP 1 on UNI 1, O EP 5 on UNI 3, and O EP 2 on ENNI. These are shown in service provider's POV in Fig. 5.11. Concatenation of OVC 1 and OVC 2 results in the EVC that subscriber ordered. Table 5.8 shows the attributes and their associated parameters that must be specified as part of the end-to-end service. Table 5.8 is organized by values that belong to the end-to-end service, that is, from subscriber's POV, values that belong to the access provider, and the values that belong to the service provider. This table is based on generic attributes and parameters that we covered in Tables 5.3–5.6.

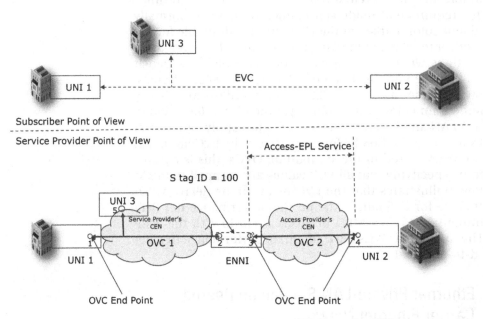

Figure 5.11 EP-LAN service based on Access-EPL for peering CENs.

In Table 5.8, row 1 shows how the EVC of MP2MP can be constructed by a service provider's MP2MP OVC and an access provider's P2P OVC. Row 2 shows the all-to-one bundling EVC attribute is "Yes." Row 3 illustrates the CE-VID to EVC map and the OVC End Point Maps with all CE-VLANs mapped to one OVC or EVC. Row 4 illustrates that the S-VID value for the OVC must agree for both networks in the ENNI end point maps. Row 5 shows how the ingress bandwidth profile for the EVC is realized in the two OVCs UNI end points. Row 6 shows that the ingress bandwidth profiles at the ENNI should be the same except reflecting a slight CIR increase due to the per-frame S-tag overhead and CM marked as color aware. Row 7 shows that CE-VLAN and CoS ID value preservation apply end to end. Row 8 states that the CoS ID for this case for the EVC has a CoS Label of M level, and in the subtending OVCs, this is accomplished by specifying that all PCP values are mapped to the M Class of Service. Row 9 illustrates that the EVC MFD performance, for CoS Label = M, and the PT of continental require MFD of 80 ms according to the MEF 23.1. The individual OVCs have adequate performance to meet the end-to-end goal. S is the subset of the ordered UNI pairs or a subset of the O EP pairs.

Table 5.8 Example of Attributes and Parameters for EP-LAN Service on Peered CENs

Row	End-to-End EP-LAN Service Related		Access-EPL Related		Service Provider Related	
	Attribute	Value	Attribute	Value	Attribute	Value
1	EVC type	Multipoint to multipoint	OVC type	Point to point	OVC type	Multipoint to multipoint
2	All-to-one bundling	Yes	N/A	N/A	N/A	N/A
3	CE-VID to EVC map	All service frames map to one EVC	OVC 2 end point map at UNI 2	All CE-VID values map to single OVC 2	OVC end point map at UNI 1 and 3	All CE-VID values map to single OVC 1
4	N/A	N/A	ENNI end point map	OVC 2 end point 3 = SVLAN ID 100	ENNI end point map	OVC 1 end point 2 = SVLAN ID 100
5	Ingress bandwidth profile per CoS identifier	CIR=100Mbps, CBS=12K, EIR, EBS=0; CM=blind; CF=0	Ingress bandwidth profile per OVC 2 EP 4 at UNI 2	CIR=100Mbps, CBS=12K, EIR, EBS=0; CM=blind; CF=0	Ingress bandwidth profile per OVC 1 EP 1 at UNI 1 and 3	CIR=100Mb/s, CBS=12K, EIR, EBS=0; CM=blind; CF=0
6	N/A		Ingress bandwidth profile per OVC 2 EP 3 at ENNI	CIR should reflect S-tag overhead, CIR, CBS, EIR and EBS same as UNI, CM=color aware, CF=0	Ingress bandwidth profile per OVC 1 EP 2 at ENNI	CIR should reflect S-tag overhead, CIR, CBS, EIR and EBS same as UNI, CM=color aware, CF=0
7	CE-VID, CoS preservation	MUST be Yes	CE-VID, CoS preservation	MUST be Yes	CE-VID, CoS preservation	MUST be Yes
8	CoS ID	Based on EVC, value=M	CoS ID based S-tag PCP	All PCP values=M	CoS ID-based S-tag PCP	All PCP values=M
9	EVC performance	MFD=80ms for label=M, PT=continental S={all ordered UNI pairs}	OVC performance	MFD=10ms; PT=regional; S={all ordered OVC EP pairs per OVC pairs}	OVC performance	MFD=50ms; PT=continental; S={all ordered OVC EP pairs per OVC pairs}

Reproduced with permission of the Metro Ethernet Forum.

5.5.3 Ethernet Private-LAN Service With Hairpin Switching by Service Provider on Peered Carrier Ethernet Networks

In case of EP-LAN service ordered by a subscriber that has two UNIs in the access provider's CEN, then an interesting situation arises where these UNIs can be connected via hairpin switching from the Service Provider. This is shown in Fig. 5.12. Here, UNI 4 is also in access provider's CEN, and so service provider buys UNI 4 and a new P2P OVC 3 from access provider. OVC 3 has an O EP 6 on UNI 4 which is connected to O EP 7 on the ENNI. This is connected to O EP 8 on ENNI with S-tag of 200. O EP 8 is made part of the MP2MP OVC 1. The connection between UNI 2 and UNI 4 is based on a hairpin switch between O EP 8 and O EP 2 on the service provider's CEN. Table 5.9 shows the attributes and values for this EP-LAN with hairpin switching at the service provider's CEN.

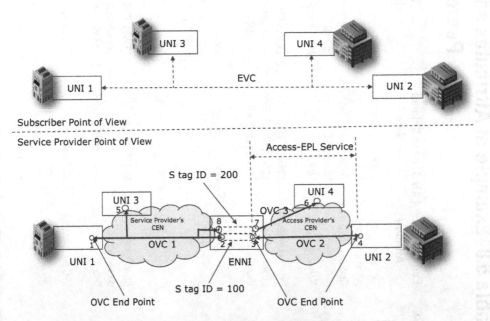

Figure 5.12 EP-LAN service with Hairpin switching for peering CENs.

Table 5.9 Attributes and Parameters for EP-LAN Service With Hairpin Switching

Row	End-to-End EP-LAN Service Related		Access-EPL Related		Service Provider Related	
	Attribute	Value	Attribute	Value	Attribute	Value
1	EVC type	Multipoint to multipoint	OVC type	Point to point	OVC type	Multipoint to multipoint
2	All-to-one bundling	Yes	N/A	N/A	N/A	N/A
3	CE-VID to EVC map	All service frames map to one EVC	OVC 2 end point map at UNI 2	All CE-VID values map to single OVC 2	OVC end point map at UNI 1 and 3	All CE-VID values map to single OVC 1
4	N/A	N/A	ENNI end point map	OVC 2 end point 3 SVLAN ID=100 OVC 3 end point 7 SVLAN ID=200	ENNI end point map	OVC 1 end point 2 SVLAN ID=100 OVC 1 end point 8 SVLAN ID=200
5	Ingress bandwidth profile per CoS identifier	CIR=100Mbps, CBS=12K, EIR, EBS=0; CM=blind; CF=0	Ingress bandwidth profile per OVC 2 EP 4 at UNI 2 and OVC 3 EP 6 at UNI 4	CIR=100Mbps, CBS=12K, EIR, EBS=0; CM=blind; CF=0	Ingress bandwidth profile per OVC 1 EP 1 at UNI 1 and EP 5 at ENNI	CIR=100Mb/s, CBS=12K, EIR, EBS=0; CM=blind; CF=0
6	N/A	N/A	Ingress bandwidth profile per OVC 2 EP 3 and OVC 3 EP 7 at ENNI	CIR should reflect S-tag overhead, CIR, CBS, EIR and EBS same as UNI, CM=color aware, CF=0	Ingress bandwidth profile per OVC 1 EP 2 and EP 8 at ENNI	CIR should reflect S-tag overhead, CIR, CBS, EIR and EBS same as UNI, CM=color aware, CF=0
7	CE-VID, CoS preservation	MUST be Yes	CE-VID, CoS preservation	MUST be Yes	CE-VID, CoS preservation	MUST be Yes
8	CoS ID	based on EVC, value=M	CoS ID based S-tag PCP	All PCP values=M	CoS ID based S-tag PCP	All PCP values=M
9	EVC performance	MFD=80 ms for label=M, PT=continental; S={all ordered UNI pairs}	OVC performance	MFD=10ms; PT=regional; S={all ordered OVC EP pairs per OVC pairs}	OVC performance	MFD=50ms; PT=continental; S={all ordered OVC EP pairs per OVC pairs}

Reproduced with permission of the Metro Ethernet Forum.

5.5.4 Ethernet Virtual Private Line Service on Peered Carrier Ethernet Networks

Fig. 5.13 shows the subscriber and service provider's POV for three instances of EVPL service offered by the service provider. Two of these three EVPL services involve using two instances of Access-EVPL service. In this example also ENNI is needed for peering of CENs belonging to service provider and access provider.

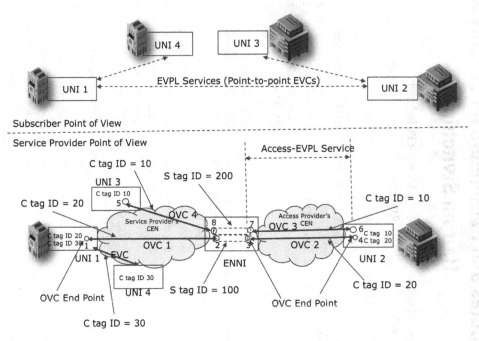

Figure 5.13 EVPL service based on Access-EVPL for peering CENs.

The subscriber orders UNI 1, UNI 2, UNI 3, and UNI 4 and three instances of EVPL service involving three P2P EVCs connecting these four UNIs from a service provider as shown in subscriber's POV. Since UNI 2 is not in the service provider's CEN, the service provider turns to an access provider and places Access-EVPL service orders. To peer the CENs of service provider and the access provider, an ENNI is also needed. The access provider provides UNI 2 and connects it to the ENNI using P2P OVC 2 with O EP 4 on UNI 2 and O EP 3 on ENNI and another OVC 3 with O EP 6 on UNI 2 and O EP 7 on ENNI. This UNI 2 has Ethernet frames with C-tag ID of 10 on OVC 3 and C-tag ID of 20 on OVC 2. The service provider in turn connects

UNI 1, and ENNI using P2P OVC 1 with O EP 1 on UNI 1 and OP E 2 on ENNI. Similarly, the service provider also connects UNI 3 with a P2P OVC 4 with O EP 5 on UNI 3 and O EP 8 on ENNI. UNI 3 has Ethernet frames with C-tag ID of 10. Also, a UNI 4 is connected by service provider with an EVC to UNI 1. UNI 1 has C-tag ID of 20 on OVC 1 and C-tag ID of 30 on EVC. These are shown in service provider's POV in Fig. 5.13. Concatenation of OVC 1 and OVC 2 results in the EVC for the EVPL service between UNI 1 and UNI 2 that subscriber ordered. Similarly, concatenation of OVC 3 and OVC 4 results in the EVPL service between UNI 3 and UNI 2. Lastly, the EVC results in the EVPL service between UNI 1 and UNI 4. Table 5.10 shows the attributes and their associated parameters that must be specified as part of the end-to-end service. Table 5.10 is organized by values that belong to the end-to-end service, that is, from subscriber's POV, values that belong to the access provider and the values that belong to the service provider.

Table 5.10 Example of Attributes and Parameters for EVPL Service on Peered CENs

Row	End-to-End EVPL Service Related		Access-EPL Related		Service Provider Related	
	Attribute	Value	Attribute	Value	Attribute	Value
1	EVC type	Point to point	OVC type	Point to point	OVC type	Point to point
2	All-to-one bundling	No	N/A		N/A	
3	CE-VID to EVC map	Map specifies what CE-VID maps to each EVC	OVC 2 end point map at UNI 2	OVC end point 6 VID 10 OVC end point 4 = CE-VID 20	OVC end point map at UNI 1 and 3	On UNI 1 OVC 1 end point 1 = CE-VID 20. On UNI 3 OVC 2 end point 5 = CE-VID 10
4	N/A	N/A	ENNI end point map	OVC 2 end point 3 = SVLAN ID 100. OVC 3 end point 7 = SVLAN ID 200	ENNI end point map	OVC 1 end point 2 = SVLAN-ID 100. OVC 4 end point 8 = SVLAN ID 200
5	Ingress bandwidth profile per CoS identifier	CIR = 100 Mbps, CBS = 12K, EIR, EBS = 0; CM = blind; CF = 0	Ingress bandwidth profile per OVC EP at UNI 2	CIR = 100 Mbps, CBS = 12K, EIR, EBS = 0; CM = blind; CF = 0	Ingress bandwidth profile per OVC EP at UNI 1 and UNI 3	CIR = 100 Mb/s, CBS = 12K, EIR, EBS = 0; CM = blind; CF = 0

Continued

Table 5.10 Example of Attributes and Parameters for EVPL Service on Peered CENs—continued

Row	End-to-End EVPL Service Related		Access-EPL Related		Service Provider Related	
	Attribute	**Value**	**Attribute**	**Value**	**Attribute**	**Value**
6	N/A	N/A	Ingress bandwidth profile per OVC EP 7 and 3 at ENNI	CIR should reflect S-tag overhead, CIR, CBS, EIR, and EBS same as UNI, CM=color aware, CF=0	Ingress bandwidth profile per OVC EP 8 and 2 at ENNI	CIR should reflect S-tag overhead, CIR, CBS, EIR, and EBS same as UNI, CM=color aware, CF=0
7	CE-VID, CoS preservation	MUST be Yes or No	CE-VID, CoS preservation	MUST be Yes	CE-VID, CoS preservation	MUST be Yes or No
8	CoS ID	Based on EVC, value=M	CoS ID–based S-tag PCP	All PCP values=M	CoS ID–based S-tag PCP	All PCP values=M
9	EVC performance	MFD=80 ms for label=M, PT=continental	OVC performance	MFD=10 ms; PT=regional	OVC performance	MFD=50 ms; PT=continental

Reproduced with permission of the Metro Ethernet Forum.

The main difference from EPL compared to EVPL in this example, is row 2, indicating all-to-one bundling is marked "No", and the corresponding change in row 3 indicating which CE-VLANs map to which EVCs. Since there are multiple OCVs mapped to one ENNI, row 4 shows that there are two S-VIDs at the ENNI, corresponding to the two OVCs in this example. It is important to note that having multiple S-tags is related to mapping of multiple OVCs to one ENNI and not due to Access-EPL or Access-EVPL service types. Additionally, row 7 shows that the end-to-end EVPL service can have CE-VID preservation as Yes or No, and since Access-EVPL is always Yes for this attribute, it will support both choices.

In all the examples of E-Access service type, we saw an attribute relating to CoS ID. This needs further explanation because when CENs are peered, service provider and access provider may not have same CoS regimes. This requires that mapping of CoS between operators has to be done correctly at the ENNI in order to deliver agreed-upon QoS as part of the SLA to the subscriber. We will discuss this in the next section.

5.6 Delivering QoS on Peering Carrier Ethernet Networks

In Section 4.4 in Chapter 4, we discussed that an SLA is an entire agreement between a subscriber and a service provider. This SLA consists of various service level objectives (SLOs), and each SLO represents a specific measureable objective. Fig. 5.14 shows a tree diagram for a typical SLA. This is a general figure to illustrate typical items covered under an SLA and then identifies QoS-related items associated with Carrier Ethernet (CE) service. Section 4.4 in Chapter 4 covered descriptions of all the components of Fig. 5.14 particularly those related to CoS, performance attributes, and policing of bandwidth profile. The important topic to cover in this section is about delivering QoS over peering CENs that requires coordination of CoS, bandwidth profile including CM and performance between service provider and access provider across ENNI so that the end customer/subscriber gets the agreed-upon QoS SLO and the overall SLA.

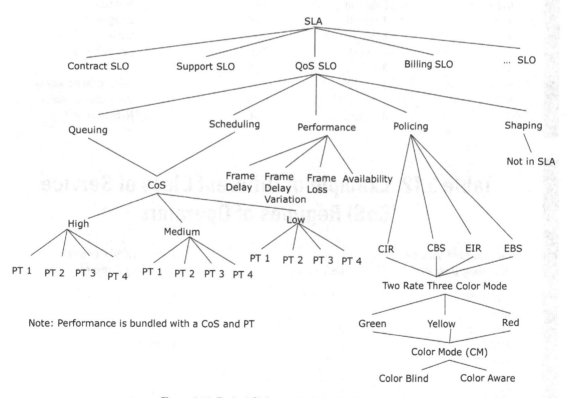

Figure 5.14 Typical SLA tree diagram for CE services.

5.6.1 CoS Coordination at External Network–Network Interface for Peering Carrier Ethernet Networks

The relationship between CoS, priority, and traffic type is shown in Table 5.11. The customer traffic type determines the priority level, and from priority level CoS or PCP value is determined; this is the value that is set in the PCP field of the C- or S-tag. Since the CENs belonging to service provider and access provider have separate administrative domain, each operator can have their own CoS regime as shown in Table 5.12 as a typical example.

Table 5.11 Relation Between Class of Service (CoS), Priority, and Traffic Type

CoS (PCP) Value	Priority	Acronym	Traffic Type
0 (default)	5 (default)	BE	Best effort
1	7 (lowest)	BK	Background
2	6 (low)	–	Spare
3	4 (better)	EE	Excellent effort
4	3	CA	Controlled Load
5	2	VI	Video, <100-ms latency
6	1	VO	Voice, <10-ms latency
7	0 (highest)	NC	Network control

Table 5.12 Example of Different Class of Service (CoS) Regimes of Operators

CoS Label of Service Providing Operator	CoS Value	CoS Label of Access Providing Operator
Super turbo	7 (highest)	Platinum
Turbo	6	Gold
Mega plus	5	Silver
Mega	4	Copper
Super	3	Tin
Extra plus	2	Rock
Plus	1	Paper
Fast	0 (lowest)	Pencil

This obviously creates a mapping issue at ENNI, so that one operator's CoS is correctly mapped to another operator's CoS. In order to facilitate mapping in a standard way, MEF 23.1[36] implementation agreement (IA) has proposed three CoS levels namely, High (H), Medium (M), and Low (L), and according to this IA, the transmitting CEN operator is responsible for mapping their internal CoS names to the MEF CoS IA Label for the Ethernet frame prior to transmitting across the ENNI, as per mutual agreement with the receiving CEN operator, so the receiving CEN can ensure compliance to the desired objectives within that CEN. This mapping to CoS IA is shown in Fig. 5.15 as an example corresponding to values shown in Table 5.12.

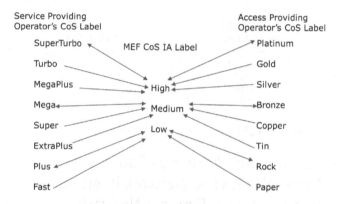

Figure 5.15 Mapping of operator CoS to MEF CoS implementation agreement.

The CoS value is specified in the PCP portion of the C-tag at the ingress UNI 1 as shown in Fig. 5.16, and if C-tag is not present, then in the PCP field of the S-tag applied at UNI 1. These CoS values are maintained in the service provider's CEN through the PBB-TE–based path specification and the associated resource reservation. At the O EP at ENNI, PBB-TE header is removed because it cannot be carried over to access provider's CEN due to that CEN being under a different administrative control. At this O EP 2, an S-tag is inserted with an S-VID that is agreed upon between service provider and access provider as per E-Access service type. Here at O EP 2 at ENNI, it becomes important to correctly map CoS value as per the MEF 23.1 IA. At O EP 3, S-tag is popped, and then access provider uses PBB-TE to switch in its CEN followed by S-tag till ingress to O EP 4 on UNI 2. When Ethernet frames travel in the reverse direction from UNI 2 to UNI 1, then correct CoS mapping must take place at O EP 3 at ENNI.

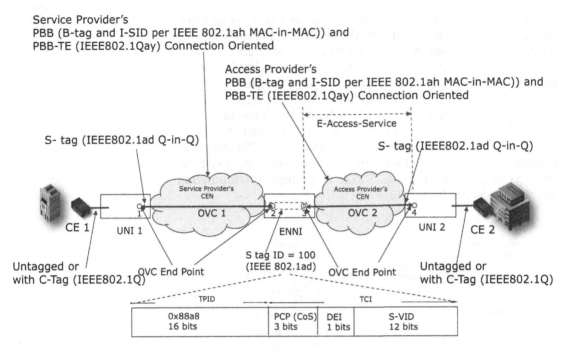

Figure 5.16 Coordination of CoS for peering CENs at ENNI.

5.6.2 Performance Attributes Coordination at External Network–Network Interface for Peering Carrier Ethernet Networks

Another important component of the QoS is performance attributes for frame delay, frame delay variation, frame loss probability, and availability. Usually once CoS and PT are chosen, then these attributes are bundled with CoS and PT as per MEF 23.1 IA as shown in Table 5.13.

Table 5.13 Performance Specifications for E-Access Service Type

CoS Label PT	CoS Label H				CoS Label M				CoS Label L			
	PT 1	PT 2	PT 3	PT 4	PT 1	PT 2	PT 3	PT 4	PT 1	PT 2	PT 3	PT 4
FD (ms)	≤10	≤25	≤77	≤230	≤20	≤75	≤115	≤250	≤37	≤125	≤230	≤390
MFD (ms)	≤7	≤18	≤70	≤200	≤13	≤30	≤80	≤220	≤28	≤50	≤125	≤240

Table 5.13 Performance Specifications for E-Access Service Type—continued

CoS Label PT	CoS Label H				CoS Label M				CoS Label L			
	PT 1	PT 2	PT 3	PT 4	PT 1	PT 2	PT 3	PT 4	PT 1	PT 2	PT 3	PT 4
IFDV (ms)	≤3	≤8	≤10	≤32	≤8 or n/s	≤40 or n/s	≤40 or n/s	≤40 or n/s	n/s	n/s	n/s	n/s
FDR (ms)	≤5	≤10	≤12	≤40	≤10 or n/s	≤50 or n/s	≤50 or n/s	≤50 or n/s	n/s	n/s	n/s	n/s
FLR (%)	≤0.01	≤0.01	≤0.025	≤0.05	≤0.01	≤0.01	≤0.025	≤0.05	n/s	n/s	n/s	n/s

Reproduced with permission from Metro Ethernet Forum.

These performance attributes are specified for OVCs, as service-level specification (SLS) attribute or OVC performance attribute. Please see Table 5.4 and examples in Section 5.5. MEF 23.2 has introduced a new PT 0.3, and we will cover that in more detail in Chapter 8.

5.6.3 Bandwidth Profile Coordination at External Network–Network Interface for Peering Carrier Ethernet Networks

The third important component of QoS is the bandwidth profile consisting of CIR, CBS, EIR, EBS, CF, and CM. Both ENNI and UNI O EPs must support CIR > 0, CBS ≥ 12,176 bytes, while the value for EIR and EBS could be zero or higher. Increments of CIR that must be supported are shown in Table 5.14.

Table 5.14 Supported CIR at UNI and ENNI for Peering CENs

Bandwidth at UNI and ENNI	Increment in CIR
10 Mbps	1 Mbps
100 Mbps	10 Mbps
1000 Mbps	100 Mbps
10 Gbps	1 Gbps

An ingress bandwidth profile can be applied per OVC end point or per CoS ID per OVC end point per ENNI. Each ingress Ethernet frame at ENNI can be subject to at most one bandwidth profile. As with bandwidth profiles at the UNI, the OVC bandwidth profiles are implemented with two-rate three-color policers that determine conformance to the profile. Green frames are forwarded and are subject to all guarantees of the service-level agreement, yellow frames are forwarded but not subject to the service-level agreements, and red frames are dropped. Ingress bandwidth profile per O EP is just like the ingress bandwidth profile per UNI. It is used to enforce a rate limit into the network over an external interface. Each ingress bandwidth profile at O EP must include suitable bandwidth profile–related parameters including CIR, CBS, EIR, EBS, CF, and CM. At the ingress O EP at ENNI, CM must be set to color aware as per MEF 26.1; however, this mandate has now been removed in MEF 26.2. The egress bandwidth profile per O EP describes egress policing by the operator on all egress Ethernet frames at ENNI that are mapped to a given O EP. An egress bandwidth profile at ENNI can be applied per O EP or per CoS ID per O EP. Each egress bandwidth profile must include suitable bandwidth profile–related parameters including CIR, CBS, EIR, EBS, CF, and CM. At the egress O EP at ENNI, CM must be set to color aware as per MEF 26.1; however, this mandate has now been removed in MEF 26.2. Note that each egress Ethernet frame at ENNI can be subjected to at most one bandwidth. Please refer to Table 5.5 and examples in Section 5.5 regarding ingress and egress bandwidth profile specifications for O EP per ENNI service attributes. Normally, no policing is needed at the egress O EP per ENNI and therefore in such cases no egress bandwidth profile per O EP need be specified.

As discussed in Section 4.2 in Chapter 4, policing of bandwidth profile is done based on "two-rate, three-color marker" model as shown in Fig. 5.14. If the bandwidth is conformant to CIR/CBS, then the frames are marked "Green" and are subject to performance as per SLS; if the bandwidth is above CIR/CBS but less than or equal to EIR/EBS, then the frames are marked "Yellow" and these frames are delivered on "best effort" basis without being subject to the SLS; and if the frames do not conform to either the CIR/CBS or the EIR/EBS, then the frames are marked "Red" and are dropped. The "two-rate, three-color marker" model is implemented by a dual-token bucket algorithm. One bucket, referred to as the "Committed" or "C" bucket, is used to determine CIR-conformant frames while a second bucket, referred to as the "Excess" or "E" bucket, is used to determine EIR-conformant, excess service frames. Each bucket is sized so as to hold respective burst sizes, CBS, for the "C" bucket and EBS for the "E" bucket.

The buckets are continuously replenished by adding tokens at the respective rates, that is, CIR/8 bytes for the "C" bucket and EIR/8 bytes for the "E" bucket. When a bucket is full, no additional tokens are added. As the service frame passes through the ingress or egress policer, the conformance is determined by comparing the number of bytes in the frame to the number of tokens in the buckets. If the "C" bucket has at least as many tokens as the number of bytes in the frame, then the frame is marked "Green," and those tokens are removed from the "C" bucket. If not, then the "E" bucket is compared. If there are enough tokens in the "E" bucket, then the frame is marked "Yellow," and those tokens are removed from the "E" bucket. Otherwise, the frame is marked "Red" and discarded. The Coupling Flag (CF) indicates whether tokens that would overflow the "C" bucket should be added to the "E" bucket. If CF = 1, then a token that would overflow the C-bucket, meaning a token that would be added except that the C-bucket is full, is now added to the "E" bucket, unless of course, it is full. If CF = 0, then the overflowing tokens are not added. CF has a negligible effect when the policing is color blind. When CF = 1 and when operating in color-aware mode, more "Yellow" service frames are allowed into the operator's network. When CF = 0, the long-term average bit rate of service frames that are declared "Yellow" is bounded by EIR. When CF = 1, the long-term average bit rate of service frames that are declared "Yellow" is bounded by CIR + EIR depending on the volume of the offered service frames that are declared "Green." In both cases, the burst size of the service frames that are declared "Yellow" is bounded by EBS.

Color mode requires bandwidth profiles to be color blind at the UNI and support color-aware mode at the ENNI (note: this mandatory requirement has now been removed in MEF 26.2 and has been made optional). Ingress bandwidth profiles per O EP at the ENNI forward Ethernet frames that are indicated as "Green" at the Committed Information Rate (CIR), and if the bandwidth profile is specified to support an Excess Information Rate (EIR) = 0, as shown in examples given in Tables 5.7–5.10, then Ethernet frames at ENNI that are indicated as "Yellow" will be dropped. The Drop Eligibility Indicator (DEI) field of the S-tag shown in Fig. 5.16 can be used advantageously to indicate color of ingress frames at the ENNI by setting DEI to 0 for "Green" and to one for "Yellow" frames, and because "Red" frames are discarded, there is no need to set any value for these "Red" frames. As per MEF 23.1[36], PCP values can also be used to identify the color of arriving frames at the ENNI. Tables 5.15 and 5.16 show values that are recommended for a three CoS label model according to MEF 23.1.

Table 5.15 CoS IA Model for Green Color Frames when Color is Marked using PCP Bits

CoS Label	PCP
H	5
M	3
L	1

Table 5.16 CoS IA Model for Yellow Color Frames when Color is Marked using PCP Bits

CoS Label	PCP
H	n/s
M	2
L	0

Bandwidth profiles at the UNI, on the other hand, must be color blind and not take account of the color of service frames when forwarding at the CIR. Color forwarding attribute related to the OVC (please see Table 5.4 and examples in Section 5.5) prevents the promotion of ingress frames mapped to an O EP from Yellow to Green. MEF 33 stipulates that the color forwarding attribute should be "Yes" for both Access-EPL and Access-EVPL services. As described before, this mandatory requirement has now been removed in MEF 26.2.

Now that we have covered the role of bridging techniques and S-tag in peering CEN, terminology and architecture of E-Access service type, Access-EPL service, Access-EVPL service, some examples of EPL and EP-LAN subscriber services utilizing E-Access service type and coordination of QoS across peering CEN including coordination of CoS, performance parameters, bandwidth profile and CM, in next section, we will cover service management in peering CEN.

5.7 Service Management in Peering Carrier Ethernet Networks

In Section 4.5 in Chapter 4, we covered that service OAM (SOAM) is essential for a carrier-grade Ethernet service. We also covered in Section 4.5 in Chapter 4 that SOAM is important because without monitoring and remediation mechanisms, there is no way to enforce SLAs. Section 4.5 in Chapter 4 also described how from an OAM perspective, there are several standards that work together in a layered fashion to provide Carrier Ethernet OAM. IEEE 802.3–2015, clause 57 (by IEEE 802.3ah amendment), defines OAM at the link level. With more of an end-to-end focus, IEEE 802.1Q (by IEEE 802.1ag amendment) defines connectivity fault management for identifying network level faults while ITU-T Y.1731 adds performance management which enables SLAs to be monitored. Taking the lead from these specifications, MEF 10.3[36] defined UNI attributes and EVC per UNI attributes to account for SOAM and MEF 30.1[36] and MEF 35.1[36] specifications provided, for CEN and peering CENs, guidelines for fault management and performance management, respectively. Now for peering CENs, MEF 33 requires that Access-EPL and Access-EVPL services must be configurable to tunnel all SOAM-based Ethernet frames at the default test and subscriber Maintenance Entity Group (MEG) levels as defined in MEF SOAM FM IA in MEF 30.1. Following the description plan of Section 4.5 in Chapter 4, we will first cover the OAM jurisdiction in peering CENs, followed by OAM framework to identify Maintenance Entity (ME), MEG, MEG End Point (MEP), and MEG Intermediate Point (MIP) and then fault management by continuity check message, link trace message (LTM), and loop-back message (LBM) in peering CENs.

Recognizing the fact that Ethernet networks often encompass multiple administrative domains, IEEE 802.1, ITU-T Study Group 13 and MEF have adopted a common, multidomain SOAM reference model. The Carrier Ethernet is divided into customer and operator maintenance levels. Fig. 5.17 shows the OAM jurisdiction defined by IEEE 802.1ah, IEEE 802.1ag, and ITU-T Y.1731 as applied to peering CENs.

Although the MEF does not require the service provider to be an operator and service provider can be an external party, it is common for the service provider to be an operator. As mentioned earlier in this book, we are assuming that the service provider is an operator. With that assumption in place, the service providing

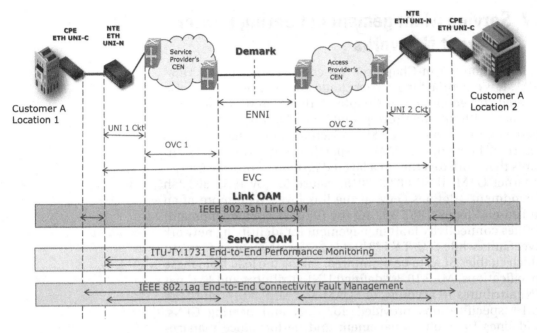

Figure 5.17 OAM jurisdiction in peering CENs.

operator has end-to-end service responsibility and responsibility for its network. In this model, as already defined in Chapter 4, an entity that requires management is called an ME. This is shown in Fig. 5.18.

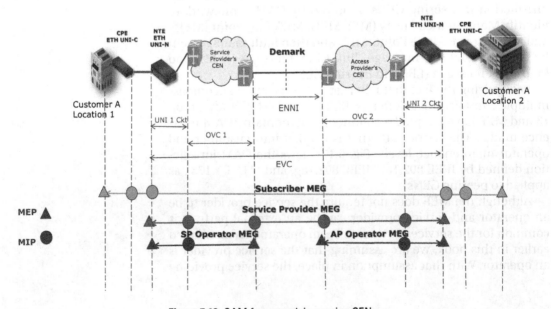

Figure 5.18 OAM framework in peering CENs.

As described earlier in Section 4.5 in Chapter 4, an ME is essentially an association between two maintenance end points within an OAM domain, where each end point corresponds to a provisioned reference point. For example, in Fig. 5.18, the arrow between the two CEs represents a subscriber ME. A MEG consists of the MEs that belong to the same service inside a common OAM domain. MEF 30.1[36] defines seven MEG levels for peering CENs as given in Table 5.17.

Table 5.17 MEG Levels for Peering CENs

MEG	Default MEG Level
Subscriber MEG	6
Test MEG	5
EVC MEG	4
Service provider MEG	3
UNI tunnel Access (UTA) MEG	3
Operator MEG	2
UNI MEG	1
ENNI MEG	1
UNI LAG link MEG	0
ENNI LAG link MEG	0

Reproduced with permission from Metro Ethernet Forum.

The MEs exist within the same administrative boundary and belong to the same P2P or multipoint EVC. For a P2P EVC/OVC, the MEG contains one single ME. A MEP is a provisioned OAM reference point which can initiate and terminate proactive OAM frames. It can also initiate and react to diagnostic OAM frames. The MEPs are indicated by triangles in Fig. 5.18. An MIP is any intermediate point in an MEG that can react to some OAM frames. An MIP does not initiate OAM frames; neither does it take action on the transit Ethernet traffic flows. The MIPs are indicated by circles in Fig. 5.18. The OAM functions are implemented into the node equipment or, if needed, are implemented in a stand-alone network demarcation device such as test heads, for monitoring specific functions such as performance management or fault management.

Performance management includes the following:
- Frame delay: Measurement of one-way and two-way (round-trip) delay from MEP to MEP.
- Interframe delay variation: Differences between consecutive frame delay measurements.

- Frame loss ratio: The number of frames delivered at an egress UNI compared to the number of transmitted frames over a specified time, for example, a month.
- Availability: Downtime is measured over, for example, a year and used to calculate the availability of the service.
  Fault Management includes the following:
- Continuity Check: "Heartbeat" messages are issued periodically by the MEPs and used to proactively detect loss of connection between end points. Continuity check is also used to detect unintended connectivity between MEGs. The continuity check is used to verify basic service connectivity and health. This is shown in Fig. 5.19.

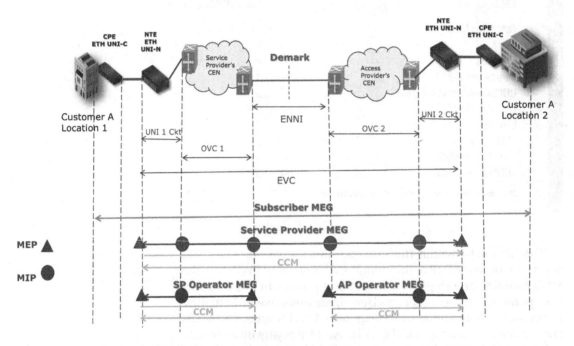

Figure 5.19 OAM fault management by continuity check message on peering CENs.

In case of fault detection, an MEP can communicate the fault by (1) remote defect indication signal where a downstream MEP detecting a fault will signal the condition to its upstream MEP(s), the behavior is similar to the RDI function in SDH/SONET networks, or (2) alarm indication signal where an MEP detecting fault can send an alarm signal to its higher level MEs, thereby informing the higher level MEs of the disruption, immediately following the detection of a fault.

- Link trace: This is an on-demand OAM function initiated in an MEP to track the path to a destination MEP. It can be thought of as a "layer 2 trace route." The procedure is similar to the loopback (LB) procedure. An MEP initiates the link trace by sending a LTM using an MAC DA and the appropriate MEG level for the ME. It is recommended that an LTM makes use of the highest CoS ID available, which will yield the lowest possible loss for a particular Ethernet service. Each MIP that belongs to this ME and the remote MEP replies by generating a Link Trace Response (LTR) unicast message. In addition to sending an LTR, each MIP also simultaneously forwards the LTM onward toward the remote MEP. If there are no faults in the path, link trace allows the originator to map the entire route to the remote MEP. If there is a fault in the path, the originator will receive LTRs from all MIPs before the fault thereby facilitating fault isolation. On the Subscriber MEG, LTM PDUs should be sent with same CE-VID that maps to the monitored EVC, that way it is guaranteed that the LTM passes the same path as the monitored EVC's service frames. It allows the transmitting node to discover connectivity data about the path. This link trace message is shown in Fig. 5.20.

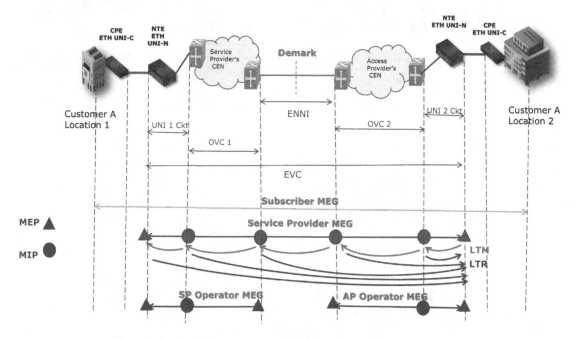

Figure 5.20 OAM fault management by link trace message on peering CENs.

- Loopback: This is an on-demand OAM function used to verify connectivity of an MEP with another MEP in the ME, but it can also be sent from an MEP to an MIP in the same ME. LBMs are defined by IEEE 802.1Q and ITU-T Y.1731 and are used on-demand as the first step to isolate a fault, which may have been detected by CCMs. An LBM is usually sent from an MEP to a remote MIP/MEP which immediately replies with LB Reply (LBR). LBM can be thought of as "layer 2 ping." LBMs are sent as an LB session of n (default is three) consecutive LBMs in a predefined time interval. If an LBR is not received within a given time interval (default is 5 s), the MEP declares the remote MEP or MIP as being faulty. The LBM uses the unicast MAC DA of the destination MIP/MEP, but it can also use multicast MAC used by CCMs. It is recommended to use the highest CoS ID for an LBM, the one that yields the lowest possible loss for that Ethernet service. An LBM is usually 64 bytes long but can be extended to any value up to the MTU size. The LBR is identical to the LBM except that the MAC DA and MAC SA are swapped. This LB message is shown in Fig. 5.21.

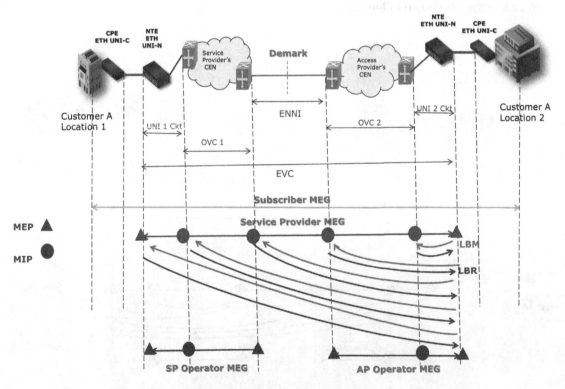

Figure 5.21 OAM fault management by Loopback message and response.

This completes the description of SOAM for peering CENs. All through this chapter we have noticed that a subscriber orders service from a service provider and because one or more subscriber locations are off-net, the service provider turns to one or more access provider(s) and orders E-Access service. This ordering process between operators is the topic of our next section and next chapter.

5.8 Need for Business-to-Business Access Service Request in Peering Carrier Ethernet Networks

Peering of CENs requires service provider placing order for Access-EPL or Access-EVPL service of the E-Access service type from one or more access providers. This order specifies the attributes and parameters that we discussed in previous sections in this chapter. This placing of order and resulting response constitutes a business-to-business (B2B) transaction which requires standardization, so that all details about location, billing, and contacts are specified in addition to the attributes and parameters of the E-Access service. This standardization helps in automating the order creation and receiving the response. This B2B transaction is called Access Service Request (ASR). There are industry organizations responsible for specifying the format of ASR forms and fields therein. Description of these ASR forms is the subject of our next chapter.

5.9 Chapter Summary

This chapter described how growth in Ethernet technology is leading to the expansion of Carrier Ethernet (CE) services beyond one operator's CEN. This expansion is also coupled with the fact that today subscribers, particularly business subscribers, have many locations that are not all in one operator's footprint. These are the reasons that are making peering of CENs necessary. To address this necessity of expanding CE services over peering CENs, MEF defined an E-Access service type which was the topic of this Chapter. In order to understand this E-Access service type, it was essential to revisit bridging techniques and tags, particularly S-tags because of the critical role S-tags play in peering CENs. This chapter explained bridging techniques and tags. The chapter then covered terminology, architecture, attributes, and parameters associated with E-Access service type to explain terms like ENNI, OVC, O EP, service provider, access provider, subscriber's point of view, service provider's point of view, and access provider's point of view. The chapter also provided reference to all IEEE, ITU-T, and MEF specifications relevant to peering CENs.

The chapter then described Access-EPL and Access-EVPL services that belong to E-Access service type and provided some examples of subscriber's services on peering CENs that use these access services. The chapter expanded on the description of delivering QoS given in Chapter 4 and described how QoS is delivered on peering CENs by coordination of CoS, performance parameters, bandwidth profile, and policing by two-rate three-color model. Description in this chapter also included SOAM over peering CENs, so that QoS is in compliance with the SLA between subscriber and service provider.

The peering of CENs requires B2B transactions between service providing operator and access providing operator. This B2B transaction, in communication industry, is called ASR, and this chapter provided a transition to ASR which is the topic of our Chapter 6.

6

STANDARDS FOR ACCESS SERVICE REQUEST

"Good order is the foundation of all good things."

Remarks by Edmund Burke.

Selling carrier Ethernet (CE) services to off-net customers over peering carrier Ethernet networks (CENs), as discussed in Chapter 5, requires service providing operators to purchase access-EPL or access-EVPL service of the E-access service type, from access providing operators. This purchase order, known as access service request (ASR), specifies the attributes and parameters that we discussed in Chapter 5. Because this placing of order constitutes a business-to-business (B2B) transaction, it requires standardization so that there is no confusion about what is being requested. The Ordering and Billing Forum (OBF) which is part of the Alliance for Telecommunications Industry Solutions[39] (ATIS) manages all the forms related to ASR and its many associated modules. These ASR forms provide a uniform means of requesting various types of services including special access or switched access services. Each request contains entries required to order particular service and establishes billing of appropriate customer accounts. It is important to note that operators have some deviations in implementing these forms, for example, some operators only entertain separate ASRs for each order, whereas other operators accept what is called as "combo" ASRs which combine multiple orders. It is, therefore, important to go to websites of operators to understand their specific ordering requirements. Detailed description, of all these forms associated with ASR and its modules, is beyond the scope of this book. However, in the next section of this chapter, we will cover a list of forms that are available and identify those forms that are relevant to E-access service type over peering CENs and provide some details of the fields in those forms. All the ASR forms are available from ATIS[39] website.

Peering Carrier Ethernet Networks. http://dx.doi.org/10.1016/B978-0-12-805319-5.00006-X

6.1 Access Service Request and its Modules

As we are describing access service request (ASR) and various modules associated with ASR, it is important to first define the term "access service." Access service means a telecommunications service and facilities associated with that service that an operator is providing to a customer. Access service may be "dedicated," in which case, it is available to a customer on a full-time, unshared basis, or it may be "switched," in which case it is available to a customer and other customers on a usage shared basis. The access service request (ASR) is generally preceded by selling activities and succeeded by a long chain of activities that includes ordering, design, order management, provisioning, installation coordination, preservice testing, service turn-up, billing, and maintenance, including ongoing coordination of testing and trouble resolution for all operator-provided facilities. The ASR forms are managed by the Ordering and Billing Forum (OBF) which is part of the Alliance for Telecommunications Industry Solutions[39] (ATIS). ATIS is an organization of the information and communications technology (ICT) companies to find solutions to their most pressing shared challenges. It has many committees and forums, and OBF is one such forum devoted to creating ordering, billing, provisioning, and exchange of information solutions about access services as well as other connectivity between telecommunications customers and providers.

It is important to note that the main module or form is called ASR, and it has many associated modules or forms; however, all these forms together are commonly known as ASR forms. The main ASR form has the following sections:
- Administrative
- Billing
- Contact
  The service-specific forms supported by the main ASR form are
- End user special access
- Feature group A
- Ring
- Switched Ethernet services
- Transport
- Trunking
- WATS access line
  There are additional service-specific forms that must accompany when ordering certain services. These additional forms are
- Additional circuit information
- Additional ring information
- End office detail
- Ethernet virtual connection
- Multi-E

- Multipoint service legs
- Network assignment information
- Ports configuration
- Service address location information
- Translation questionnaire
- Virtual connection (request types V or X)
- Virtual concatenation (request types S, E, R, V, X)

In short, an access service request to an operator must accompany ASR form along with services specific form(s) and, if applicable, some additional forms. These forms can be obtained from ATIS.[39] Table 6.1 provides some high-level description of various ASR forms.

Table 6.1 Brief Description of Various Access Service Request Forms

Form Name	Type	Description	Sections in the Form
ACI (additional circuit information)	Additional	Used when the customer or the provider is stipulating circuit-specific information that cannot readily be provided on a service-specific request form.	• Circuit detail
ARI (additional ring information)	Additional	Used to order additional ring segments, specify ordering options, and describe the termination of the ARI.	• Circuit detail • Primary location • Secondary location • Remarks
ASR (access service request)	Main module	Used to request various services as specified in contracts and tariffs. Contains all information required for administrative, billing, and contact details. Must always be associated with a service-specific form containing circuit and location detail necessary for designing and provisioning the request.	• Administrative • Bill • Contact
EOD (end office detail)	Additional	Used to forecast traffic routed from end offices subtending a tandem, identify end offices for SAC code activity, identify subtending end offices for originating traffic requests. Estimate traffic distribution requirements when traffic is switched through an access tandem.	• Administrative • Remarks

Continued

Table 6.1 Brief Description of Various Access Service Request Forms—continued

Form Name	Type	Description	Sections in the Form
EUSA (end user special access)	Service specific	Used to order a special access service when neither end is an access customer terminal location (ACTL).	• Circuit detail • Primary location • Secondary location • Remarks
EVC (Ethernet virtual connection)	Additional	Used for providing circuit details and remarks/details related to EVC requests. This form is needed when ordering EUSA or SES services.	• Circuit detail
FGA (feature group A)	Service specific	Used to provide information about lineside access to provider end-office switches with an associated seven-digit local telephone number.	• FGA service detail section • FGA service option section • FGA location section • Remarks
MSL (multipoint service legs)	Additional	Used to order a bridge circuit configuration. The configuration may be either a leg to an end user location off a bridge or another bridge off a bridge.	• Circuit detail • Remarks
MULTIEC (multiex-change company)	Additional	Used to order access services when more than one provider is involved in provisioning the services. Provides the additional administrative and billing detail information for each provider involved.	• Access service coordination • Company detail • Other exchange Company detail
NAI (network assignment information)	Additional	Used to provide immediate connecting facility assignments, an alternate facility/ACTL, and drop port equipment assignment information.	• Alternate service detail • Circuit detail
PC (ports configuration)	Additional	Used to request specific equipment configu-rations, such as SONET/DWDM.	• Configuration
RING (Ring)	Service specific	Used to define the service and first segment of a ring configuration.	• Circuit detail • Primary location • Secondary location • Remarks
SES (switched Ethernet service)	Service specific	Used to request switched Ethernet service	• Circuit detail • Address detail • Remarks

Table 6.1 Brief Description of Various Access Service Request Forms—continued

Form Name	Type	Description	Sections in the Form
SALI (service address location information)	Additional	Used for end user locations that are service addresses.	• Address detail
TQ (translation questionnaire)	Additional	Used to identify which and how CICs, SACs, and NXXs will be routed by the trunks being ordered. Use when local, feature group B (FGB), or feature group D (FGD) translations/routing are required for new, change, disconnect, or activation activities.	• Common • Matrix • Interconnection/STP translation routing • Feature group D • Remarks • SAC NXX code activity • Remarks
Transport	Service specific	Used to provide all the information for ordering services between an access customer terminal location (ACTL) and an end user location such as special access service, switched access facility, or unbundled transport.	• Circuit detail • Remarks
Trunking	Service specific	Used to order trunk side BSA, feature group B, C, D, local interconnection, wireless trunking, CCS links and unbundled STP ports.	• Service details • Remarks
TSR (testing service request)	Service specific	Used to order various testing services in the access tariffs.	• Administrative • Testing • Remarks
VC (virtual connection)	Additional	Used for providing circuit details and remarks details related to VC requests.	• Circuit detail
WATS (Wats)	Service specific	Used to order WATS access lines.	• WATS circuit detail section • Remarks

For ordering E-Access service type over peering CENs, the service providing operator needs to only use the forms shown in Table 6.2, to place an Access-EPL or an Access-EVPL service from an access providing operator. As mentioned before, it is important to note that operators have some deviations in implementing these forms. It is, therefore, important to go to Websites of operators to understand their specific ordering requirements.

Table 6.2 List of Forms Needed for E-Access Service Type for Peering CENs

Order Type	ASR	EUSA	SALI	Transport	EVC
Access-EPL	✓	✓	✓	—	✓
Access-EVPL	✓	✓	✓	—	✓
ENNI transport	✓	—	✓	✓	—

Now that we have explored various OBF forms and identified specific forms that are needed for E-access service using peering CENs; in the next section, we will explore fields in these applicable forms in some detail. Please note that the actual forms can be obtained from ATIS.[39]

6.2 ASR Forms for E-Access Service for Peering CENs

From Section 5.4 of Chapter 5, we know that MEF 33 defines two Ethernet services for the E-access service type namely, access-Ethernet private line or access-EPL service which are "private" or port-based services and access-Ethernet virtual private line or access-EVPL service where service are "virtual private" or VLAN-based services. Fig. 6.1 shows the schematic of an access-EPL service. As described with reference to Fig. 5.10 in Section 5.5 of Chapter 5, because UNI 2 is not in the service provider's CEN, the service provider turns to an access provider and places an access-EPL service

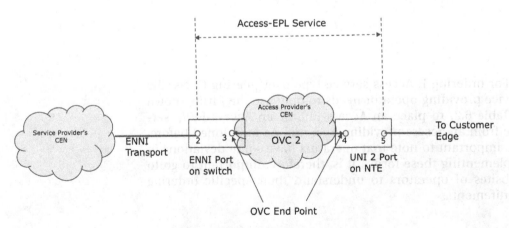

Figure 6.1 Access-EPL service.

order over the peering CENs of service provider and the access provider. The access providing operator provides UNI 2 and an ENNI port and also an OVC2 connecting OVC end point 4 on UNI 2 with OVC end point 3 on ENNI, completing the access-EPL service.

This access-EPL service, in this example, includes an ENNI port on a network switch, UNI port on an NTE, and an OVC connecting an OVC end point on UNI to an OVC end point on ENNI.

Fig. 6.2 shows the schematic of an access-EVPL service. As described with reference to Fig. 5.13 in Section 5.5 of Chapter 5, the subscriber, in this example, orders UNI 1, UNI 2, UNI 3, and UNI 4 and an EVPL service involving three point-to-point EVCs connecting these four UNIs from a service provider. As UNI 2 is not in the service provider's CEN, the service provider turns to an access provider and places an access-EVPL service order to peer

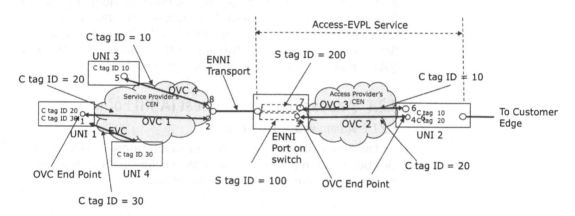

Figure 6.2 Access-EVPL service.

the CENs of service provider and the access provider. The access provider provides UNI 2 and connects it to the ENNI using point-to-point OVC 2 with O EP four on UNI 2 and O EP 3 on ENNI and another OVC 3 with O EP 6 on UNI 2 and O EP 7 on ENNI. This UNI 2 has Ethernet frames with C tag ID of 10 on OVC 3 and C tag ID of 20 on OVC 2. The service provider in turn connects UNI 1 and ENNI using point-to-point OVC 1 with O EP 1 on UNI 1 and OP E 2 on ENNI port on its CEN. Similarly, the service provider also connects UNI 3 with a point-to-point OVC 4 with O EP 5 on UNI 3 and O EP 8 on ENNI. UNI 3 has Ethernet frames with C tag ID of 10. Also, UNI 4 is connected by service provider with an EVC to UNI 1. UNI 1 has C tag ID of 20 on OVC 1 and C tag ID of 30 on EVC. Concatenation of OVC 1 and OVC 2 results in the EVC for the EVPL service between UNI 1 and UNI 2 that subscriber ordered. Similarly concatenation of OVC 3 and OVC 4 results in the EVPL service between UNI 3 and UNI 2. Finally the EVC results in the

EVPL service between UNI 1 and UNI 4. Like the access-EPL service shown in Fig. 6.1, this access-EVPL service, shown in Fig. 6.2, also includes an ENNI port on a network switch, UNI port on an NTE, and OVC connecting an OVC end point on UNI to an OVC end point on ENNI. The difference here is that not all of the service frames at the UNI 2 are mapped to one OVC but are mapped to different members of a set of OVC end points belonging to different OVCs based on CE-VLAN ID or C tags, and therefore services are multiplexed at both UNI as well as ENNI ports.

As shown in Figs. 6.1 and 6.2, access-EPL and access-EVPL services consist of ordering ENNI port, UNI port, and OVC. The sequence of order is important. ENNI port is ordered first, then UNI is ordered, and finally OVC is ordered because OVC order has to reference both ENNI and UNI circuits. Ordering ENNI transport is optional and can be ordered by a service providing operator if needed. With this background on access-EPL and access-EVPL services and with reference to Table 6.2, the brief descriptions of fields, available in public domain, in the ASR, EUSA, SALI, EVC (OVC), and ENNI transport forms are given below.

6.2.1 Access Service Request (ASR) Form

6.2.1.1 Administration Section

The Administrative section contains information about the service being ordered. The descriptions of fields on the ASR administrative section are given in Table 6.3 below.

Table 6.3 Administrative Section in ASR Form

Field	Description	Populate With/Usage
CCNA	Customer carrier name abbreviation—(3 alpha characters)	Required
PON	Purchase order number—identifies the customer's unique purchase order number for the request. (16 alpha/numeric characters)	Required
VER	Version identification—identifies the provider service center (4 alpha/numeric characters)	Required Initial ASR request=00; if sending
ICSC	Interexchange customer service center—identifies the provider service center (4 alpha/numeric characters)	Required
D/TSENT	Date and time sent—identifies the date and time that the ASR is sent by the customer. (17 alpha/numeric characters including three hyphens)	Required—system populated

Table 6.3 Administrative Section in ASR Form—continued

Field	Description	Populate With/Usage
CBD	Call before dispatch—identifies site operational availability in hours.	Optional
DDD	Desired due date—identifies the customer's desired due date (10 alpha/numeric characters).	Required Valid entry: YYYYMMDD standard intervals: ICB, capacity needed
PROJECT	Project identification—identifies the project which the request is to be associated (16 alpha/numeric characters)	Optional Note: Examples of the use of this field would be relating multiple access service requests, previously negotiated orders, and so forth
CNO (case number)	Project identification—identifies the quotation tracking number assigned by the provider in response to a provisioning arrangement inquiry, e.g., diversity (12 alpha/numeric characters).	Conditional Valid entry: Customer to enter the quote ID provided for ICB based pricing. Note: For non-MSA products
REQTYP	Requisition type and status—identifies the type of service being requested and the status of the request (2 alpha characters).	Required Valid entries: REQTYP=SD (special Access) REQTYP=ED (end-user request)
ACT	Activity—identifies the activity involved in the service request (1 alpha character)	Required Valid entries: N=new D=Disconnection C=change or Modify existing service
RTR	Response type requested—identifies the type of confirmation response requested by the customer. (2 alpha/numeric character).	Required Valid entry: F=send FOC only
QSA	Quantity service address location information—identifies the total number of service address location information forms being sent by the customer. (2 numeric characters)	Required Example: 01 Note: Prohibited when EVCI=A, leave field blank.
EVCI	Ethernet virtual connection Indicator – identifies that an Ethernet virtual connection Form is associated with this service request (1 alpha character)	Required Valid entries: A=EVC/OVC form attached

Continued

Table 6.3 Administrative Section in ASR Form—continued

Field	Description	Populate With/Usage
CUST	Customer name—identifies the name of the customer who originated this request.	Required
CKR	Customer circuit reference—identifies the circuit number or range of circuit numbers used by the customer.	Required Valid entry: C
UNIT	Unit identification—identifies whether the quantity (QTY) field contains number of circuits, ring segments, busy hour minutes of capacity (BHMC) for switched access service or percent of market share.	Required
PIU	Percentage of interstate usage—identifies the expected interstate usage for the access service for the request (3 alpha/numeric characters)	Required Valid entry: 100
QTY	Quantity—identifies the quantity involved in the service request	Required Example: 1
SPEC	Service and Product Enhancement Code—identifies a specific product or service offering. (5 alpha/numeric characters, 7 alpha/numeric characters)	Required
BAN	Billing account number—identifies the billing account to which the recurring and nonrecurring charges for this request will be billed.	Required
ACTL	Access customer terminal location—identifies the CLLI code of the customer facility terminal location. (11 alpha/numeric characters)	Required
RPON	Related purchase order number—identifies the PON of a related access service request. RPON and remarks should be used to associate new ports to an existing service.	Optional
REMARKS	Remarks—identifies a free flowing field, which can be used to expand upon and clarify other data on this request. (186 alpha/numeric characters)	Required Valid entry: Additional clarifying information

6.2.1.2 Billing Section

The billing section in ASR form is required to insure the accuracy of the billing data. These fields are order-specific and are not required on subsequent revisions unless changes are being made to what was originally requested. Descriptions of fields in the billing section are given in Table 6.4.

Table 6.4 Billing Section in ASR Form

Field	Description	Populate With/Usage
BILLNM	Billing name—identifies the name of the person, office, or company to whom the customer has designated that the bill be sent.	Required
SBILLNM	Secondary billing name—identifies the name of a department or group within the designated BILLNM entry (25 alpha/numeric characters)	Optional
ACNA	Access customer name abbreviation—identifies the common language code for the customer who should receive the bill for the ordered service (3 alpha characters)	Required
TE	Tax exemption—identifies that the customer has submitted a tax exemption form to the provider.	Required
FUSF	Federal universal service fee—identifies the service being ordered on this request should be either assessed or exempted from the Federal universal service fee (FUSF)	Required
STREET	Street address (Bill)—identifies the street of the billing address associated with the billing name. (25 alpha/numeric characters)	Required
ROOM	Room (Bill)—identifies the room for the billing address associated with the billing name (6 alpha/numeric characters)	Required
CITY	City (Bill)—identifies the city, village, township of the billing address associated with the billing name (25 alpha/numeric characters)	Required
STATE	State (Bill)—identifies the two character postal code for the state of the billing address associated with the billing name (2 alpha characters)	Required
ZIPCODE	Zipcode (Bill)—(12 alpha/numeric character)	Required
BILLCON	Billing contact (Bill)—(15 alpha/numeric characters)	Required
TEL NO	Telephone number (Bill) contact—(17 numeric characters, including three preprinted hyphens)	Required
BILLCON EMAIL	Billing contact electronic mail address (Bill)—identifies the electronic mail address of the billing contact when a customer profile does not already exist.	Required

6.2.1.3 Contact Section

The contact section of the ASR form contains the relevant contact information for the person initiating the order. Details of fields in this section are given in Table 6.5.

Table 6.5 Contact Section in ASR Form

Field	Description	Populate With/Usage
INIT	Initiator—identifies the customer's employee who originated the request (15 alpha/numeric characters)	Required
TEL NO	Telephone number (INIT)—identifies the telephone number of the number of customer's employee who originated the request (17 numeric characters, including 3 preprinted hyphens)	Required
INIT FAX NO	Initiator fax number (INIT)—identifies the fax number of the initiator (12 numeric characters, including 2 preprinted hyphens)	Required
INIT EMAIL	Initiator electronic mail address (INIT)—identifies the electronic mail address of the Initiator. (60 alpha/number characters)	Required
DSGCON	Design/engineering contact information—identifies the employee of the customer or agent who should be contacted on the design/engineering matters and to whom the DLR may be sent. (15 alpha numeric characters)	Required
TEL NO	Telephone number (DSGCON)—identifies the telephone number of the customer's employee who should be contacted on design/engineering matters. (17 numeric characters, including 3 preprinted hyphens)	Required
DSG FAX NO	Design Fax Number (DSGCON)—identifies the fax number on the design contact. (12 Numeric characters- including 2 preprinted hyphens)	Required
DSG EMAIL	Electronic mail address (DSGCON)—identifies the electronic mail address of the DSGCON (60 alpha/numeric characters)	Required
STREET	Street address (DGSCON)—identifies the street address for the design/engineering contact (25 alpha/numeric character)	Required
FL	Floor (DSGCON)—identifies the floor of the design/engineering contact's address. (3 alpha/numeric characters)	Required
CITY	City (DSGCON)—identifies the city of the design/engineering contact's address (25 alpha/numeric characters)	Required
STATE	State (DSGCON)—identifies the two character postal code for the state/province of the design/engineering contact's location	Required
ZIPCODE	Zip code (DSGCON)—identifies the zip code or postal code of the design/engineering contact's address	Required
IMPCON	Implementation contact—identifies the Customer's employee who is responsible for control of installation and completion. (15 alpha/numeric characters)	Required
TEL NO	Telephone number (IMPCON)—identifies the telephone number of the implementation contact (17 numeric characters, including 3 preprinted hyphens)	Required

6.2.2 SALI (Service Address Location Information) Form

The Service Location Information Form (SALI) must be provided to accurately reflect the service address information. Details of fields in this form are shown in Table 6.6 below.

Table 6.6 Service Address Location Information (SALI) Form

Field	Description	Populate With/Usage
PI	Primary location indicator—identifies that the service address location information being provided is a primary location	Conditional—required when request type begins with E and location is the primary. Should not be used for the end user (Z) location. Valid entry: Y = Yes
EUNAME	End user name—identifies the end user name associated with the termination location (25 alpha/numeric characters)	Required Example: XYZ, Inc
NCON	New construction—identifies that the service address is a new construction (1 alpha character)	Optional Valid entry: Y = Yes
SAPR	Address number prefix—identifies the prefix for the address number of the service address (6 alpha/numeric characters)	Optional
SANO	Address number—identifies the number of the service address (10 alpha/numeric characters)	Required
SASF	Address number suffix—identifies the prefix for the address number of the service address (4 alpha/numeric characters)	Optional
SASD	Street directional prefix—identifies the street directional prefix for the service address (2 alpha characters)	Optional Valid entry: N, S, E, W, NE, NW, SE, SW
SASN	Street name—identifies the street name of the service address (60 alpha/numeric characters)	Required
SATH	Street type—identifies the thoroughfare portion of the street name of the service address (7 alpha/numeric characters)	Required
LD1, LD2, LD3	Location designators—identifies additional specific information related to the service address (e.g., building, floor, room; 4 alpha characters)	Optional

Continued

Table 6.6 Service Address Location Information (SALI) Form—continued

Field	Description	Populate With/Usage
LV1, LV2, LV3	Location values—identifies the value associated with the first location designator of the service address (10 alpha/numeric characters)	Optional
CITY	City—identifies the city, village, township, and so forth of the service address (32 alpha/numeric characters)	Required
STATE	State—identifies the state/province of the service address (2 alpha characters)	Required
ZIP	ZIP/postal code—identifies the ZIP code, ZIP code + extension, or postal code of the service address (12 alpha/numeric characters)	Required
AAI	Additional address information—identifies additional location information about the service address (150 alpha/numeric characters)	Optional Examples: specific access hours, availability, contact notification requirements, specific access requirements
JS	Jack status—indicates whether the access service is to terminate at a new or existing registered jack or demarcation. (1 alpha character)	Required—when ACT type is N Valid entry: F
LCON	Local contact—identifies the local contact name for access. (15 alpha/numeric characters)	Required
ACTEL	Access telephone number—identifies the telephone number to be used for the purpose of arranging access to the service address location for installation purposes [14 numeric characters (excluding 3 preprinted hyphens)]	Required
LCON EMAIL	Local contact email address—identifies the electronic mail address of the local contact (60 alpha/numeric characters)	Required

6.2.3 EUSA (End User Special Access) Form

EUSA form is for ordering UNI and ENNI ports or channels or interfaces. This form has three sections namely, circuit details section, primary location section and secondary location section. Details of fields are given in Table 6.7.

Table 6.7 End-User Special Access (EUSA) Form

Field	Description	Populate With/Usage
NC	Network channel—identifies the network channel (NC) code for the circuit (s) involved. The NC code describes the channel provided by the operator from the end user's location. Four alpha/numeric character. The NC also describes portions of a circuit: • ACTL to HUB • HUB to HUB • HUB to end user location	Required
NCI	Network channel interface code—identifies the electrical conditions on the circuit at the primary location. Five alpha/numeric characters minimum, and 12 Alpha/numeric maximum	Required
SECNCI	Secondary network channel interface—identifies the electrical conditions on the circuit at the secondary ACTL or end-user location. Five alpha/numeric characters minimum, and 12 alpha/numeric maximum.	Required
ESP	Ethernet service point—identifies the Ethernet switching point, terminating equipment or terminating location, in CLLI code format, at the UNI/ENNI termination.	Required
PRILOC	Primary location—identifies the primary end of the circuit being provided. Twelve alpha/numeric character	Required
S25	Surcharge status—identifies whether a surcharge is applicable (nonexempt) or nonapplicable (exempt) for the number of circuits ordered between two customer locations. Eight alpha/numeric character	Required
ER	Exempt reason—tells the provider why a circuit is exempt from the special access surcharge. One numeric character	Required
GETO	General exchange tariff options code—identifies the requirement for nontariff or secondary tariff options in conjunction with the access service and special arrangements (third party billing). One alpha characters	Optional Valid entries: • Blank=No option • W=provide inside wiring and bill the customer.
SECLOC	Secondary location—identifies the terminating end of the circuit, a provider end office, or first point of switching for the circuit being provided. Eleven alpha/numeric characters	Required Populate with the 11-digit CLLI for the associated ENNI location.
SEI	Switched Ethernet indicator—identifies this request is ordering UNI/ENNI interface to provider-owned Ethernet switch/router. One alpha characters	Optional Valid entry: Y=ordering a UNI/ENNI Ethernet switch

Continued

Table 6.7 End-User Special Access (EUSA) Form—continued

Field	Description	Populate With/Usage
S25	Surcharge status—identifies whether a surcharge is applicable (nonexempt) or nonapplicable (exempt) for the number of circuits ordered between two customer locations. Eight alpha/numeric character	Required
ER	Exempt reason—tells the provider why a circuit is exempt from the special access surcharge. One numeric character	Required
REMARKS	Remarks—identifies a free-flowing field which can be used to expand upon and clarify other date on this form.	Optional

6.2.4 OVC (EVC) Form

The operator virtual connection (OVC) form is exactly same as the Ethernet Virtual Connection form (EVC) only the populated fields are different, and this form must be completed to provide specific details in regard to the ordering and provisioning of the "OVC." Descriptions of the fields in this form are given in Table 6.8.

Table 6.8 Operator Virtual Connection (OVC) Form

Field	Description	Populate With/Usage
CCNA	Customer carrier name Abbreviation—(3 alpha characters)	Required
PON	Purchase order number—identifies the customer's unique purchase order number for the request (16 alpha/numeric characters)	Required
VER	Version identification—identifies the customer version number	Required Initial ASR request=00; If sending SUPP then assign next number (example: 01 then 02 then 03)
ASR NO	Access service request number—identifies the number that may be generated by the provider to identify a customer's request.	Optional
EVC NUM	Ethernet virtual connection reference number—identifies a unique number associated with the Ethernet virtual connection. The EVC number is customer assigned and is returned on the confirmation notice to the ordering customer.	Required

Table 6.8 Operator Virtual Connection (OVC) Form—continued

Field	Description	Populate With/Usage
NC	Network channel—identifies the network channel (NC) code for the circuit (s) involved. The NC code describes the channel provided by operator from the end user's location. Four (4) alpha/numeric character. The NC also describes portions of a circuit: • ACTL to HUB • HUB to HUB • HUB to end user location	Required
EVCID	Ethernet virtual connection identifier—identifies the provider-assigned EVC identifier	Optional 28 alpha/numeric characters
NUT	Number of UNI/ENNI terminations—reflects the number of UNI/ENNI termination occurrences being affected by the service request	Conditional Valid entry: 01–20
SVP	SVLAN ID preservation—identifies that the customer is requesting SVLAN ID preservation on a requested OVC	Conditional
EVCCKR	Ethernet virtual connection customer circuit reference—identifies the circuit number used by the customer	Optional
UREF	User network interface (UNI) reference number—identifies the reference number associated to the UNI port or ENNI termination point to which the EVC/OVC mapping requirements will be applied	Conditional
EI	ENNI indicator—identifies when the UREF is an ENNI	Conditional
AUNT	Associated UNI/ENNI termination—identifies the UREF termination point associated with the physical port being requested on the ASR	Conditional
UACT	User network interface (UNI) activity indicator—identifies the activity that is taking place at this UNI/ENNI termination point	Conditional
RPON	Related purchase order number—identifies the PON of a related access service request	Conditional
NCI	Network channel interface code—identifies the electrical conditions on the circuit at the primary location. Five alpha/numeric characters minimum, and 12 alpha/numeric maximum	Required See Service Codes section
L2CP	Layer 2 control protocol—identifies a set of protocols that are used for various control purposes that allow the Ethernet network to effectively process information for subscribers who chose to deploy 802.1Q bridges.	Conditional

Continued

Table 6.8 Operator Virtual Connection (OVC) Form—continued

Field	Description	Populate With/Usage
EVCSP	Ethernet virtual connection switch point—identifies the Ethernet switching point, CLLI code format, at the UNI/ENNI termination	Conditional
RUID	Related UNI identifier—Identifies the provider's related circuit ID for a UNI or ENNI against which the OVC/EVC is requested. Twenty-eight alpha/numeric characters. NOTE 1: EVC/OVC form should reflect both the ENNI and UNI ECCKT ID's. NOTE 2: For new EVC/OVC orders, populate with the associated UNI/ENNI circuit ID.	Required The ENNI circuit must be ordered before any UNI orders being submitted. In addition, the UNI must be ordered before the OVC/EVC because the OVC order must reference the appropriate ENNI and UNI.
R/L	ROOT/LEAF—indicates that the UNI is either a root or a leaf in a rooted multipoint EVC.	Conditional
SVACT	Service virtual local area network activity—identifies the activity requested for the SVLAN	Conditional
SVLAN	Service virtual local area network—the identifier found within the service tag which is typically associated with the OVC end points at an ENNI.	Conditional
EVCMPID	EVC meet point ID—specifies the physical facility ID interconnecting the two service providers in an EVC meet-point configuration.	Conditional
OTC	Other exchange company—identifies the EC or company code of the network facing switch of the provider in an EVC meet-point service arrangement	Conditional
ASN	Autonomous system number—indicates the unique number identifying the customer Internet network ordering the BGP service.	Conditional
VPN-ACT	Virtual private network identifier activity—identifies the activity requested for the VPN-ID	Conditional
VPN-ID	Virtual private network identifier—indicates a unique identifier for the virtual private network that creates a secure network connection over a public network.	Conditional
VACT	Customer edge virtual local area network activity indicator—identifies the activity requested for the CE-VLAN	Conditional
CE-VLAN	Customer edge virtual local area network indicator—an identifier derivable from a content of a service frame that allows the service frame to be associated with an EVC/OVC at the UNI.	Conditional

Table 6.8 Operator Virtual Connection (OVC) Form—continued

Field	Description	Populate With/Usage
LREF	Level of service reference number—identifies the reference number associated with the level of service mapping configuration being requested	Conditional
LOSACT	Level of service activity indicator—identifies the activity for the level of service at this UNI termination occurrence.	Conditional Valid entries C = change D = disconnect K = cancel N = new
LOS	Level of service name—identifies a name for a provider-defined level of service performance associated with the Ethernet product offering.	Required Valid entries Standard data
SPEC	Service and product enhancement code—identifies a specific product or service offering	Conditional
P-BIT	Priority bit—an optional parameter within the Ethernet frame to specify priority. In this application, it will be used to map certain traffic to a given level of service on an EVC/OVC, when the provider supports multiple level of service per EVC/OVC	Conditional
CMI-I	Color mode indicator—"E" for enable, "D" disable	Conditional
BDW	Bandwidth—identifies the average rate in bits per second of ingress service frames up to which the network delivers frames and meets the performance objectives defined by the LOS service attribute	Conditional
DSCP	Differentiated services code point—identifies an integer value encoded in the DiffServ field of an IP header	Conditional
TOS	Type of service—identifies the quality of service desired	Conditional
Remarks	Remarks—identifies a free-flowing field which can be used to expand upon and clarify other date on this form. Up to 125 alpha/numeric characters	Optional

6.2.5 Transport (ENNI) Form

This form is required for transport services that terminate to a common carrier POP. The descriptions of fields in this form are given in Table 6.9.

Table 6.9 Transport (for ENNI) Form

Field	Description	Populate With/Usage
NC	Network channel—identifies the network channel (NC) code for the circuit (s) involved. The NC code describes the channel provided by operator from the end user's location. Four alpha/numeric The NC also describes portions of a circuit: • ACTL to HUB • HUB to HUB • HUB to end user's location	Required
NCI	Network channel interface Code—identifies the electrical conditions on the circuit at the primary location. Five alpha/numeric characters minimum, and 12 alpha/numeric maximum	Required
SECNCI	Secondary network channel interface—identifies the electrical conditions on the circuit at the secondary ACTL or end user location. Five alpha/numeric characters minimum, and 12 alpha/numeric maximum	Required
CFA	Connecting facility assignment—identifies the provider carrier system and channel to be used from a wideband analog, high capacity, or optical network facility when the customer has assignment control. 42 alpha/numeric characters	Conditional
MUX LOC	Multiplexing location—identifies the CLLI code of the provider central office which provides multiplexing for a service riding a high-capacity service. Eight alpha/numeric characters	Conditional
SCFA	Secondary connecting facility assignment—identifies the provider carrier system and channel to be used from a wideband analog, high capacity or optical network facility for a thru-connect configuration when the customer has assignment control. 42 alpha/numeric characters	Conditional
GETO	General exchange tariff options code—identifies the requirement for nontariff or secondary tariff options in conjunction with the access service and special arrangements (third-party billing). One alpha characters	Optional Valid entries: • Blank = No option • Y = provide inside wiring and bill end user customer directly.
SECLOC	Secondary location—identifies the terminating end of the circuit, a provider end office or first point of switching for the circuit being provided. Eleven alpha/numeric characters	Required Populate with the 11-digit CLLI for the associated ENNI location.
SR	Special routing code—identifies the type of special routing requested. Three alpha/numeric characters	Conditional
SEI	Switched Ethernet indicator—identifies this request is ordering UNI/ENNI interface to provider owned Ethernet switch/router. One alpha characters	Optional Valid entry: Y = ordering a UNI/ENNI Ethernet switch
Remarks	Remarks—identifies a free-flowing field which can be used to expand upon and clarify other date on this form. Up to 125 alpha/numeric characters	Optional

In the descriptions of fields of various forms shown in Tables 6.3–6.9, we came across some terms like NC, NCI, SECNCI, CLLI codes and so forth. These terms are called Common Language codes and are trademarks of Telcordia Technologies, Inc., doing business as iConnective.[40] Common Language Information Services provided by iConnective include CLLI Codes, CLFI Codes, CLCI Codes, NC/NCI Codes, CLEI Codes, USOC Codes, and FID Codes. An example of NC/NCI/SECNCI code are shown in Table 6.10.

Table 6.10 An Example of Common Language Codes Used in ASR Forms

NC	Location	NCI	SECNCI
KSE1	AC/EU to CO	02LNF.A02	02CXF.10G

In this example shown in Table 6.10, first two characters of NC code specify port bandwidth. In this example, first two characters are "KS," signifying 10 Gbps port. The third character in NC code specifies if it is port-based or VLAN-based connection. In this example, the third character is "E," signifying that it is VLAN-based connection. Finally, the fourth character in the NC code specifies CIR which, in this example is "1" signifying a CIR of 1000 Mbps. In the location column of Table 6.10, AC means access customer's POP, CO is for central office, and EU means end User. NCI and SECNCI codes specify type of physical interfaces at primary and secondary locations. In NCI code of "02LNF.A02," 02 means two conductors (2-fiber strands), LN means local area network interface, F means fiber, A02 means 1310 nm, single-mode fiber (10,000 Base ZX). In SECNCI code of "02CXF.10G," 02 means two conductors (2-fiber strands), CX means digital termination on a switch in CO, F means fiber, and 10G means 10 Gigabit Ethernet termination to CO.

Based on all these descriptions, it is clearly evident that placing orders using ASR forms requires detailed knowledge of not only the forms but also of the Common Language codes. Table 6.11 below lists the applicable tables from Chapters 4 and 5 for attributes and parameters for E-access service type specified by MEF 26.1 and MEF 33, and the OBF-specified ASR forms that we covered in this chapter for E-access service type that a service provider has to send to an access provider. It is important to recall

Table 6.11 MEF and ASR Specific Tables Related to E-Access Service Type

Attributes and Parameters for E-Access Service Type Specified by MEF 26.1 and MEF 33		OBF Specified ASR Forms Relevant to E-Access Service Type	
Item	Table in Chapters 4 and 5	Item	Table in Chapter 6
ENNI	Table 5.3	ASR	Tables 6.3, 6.4 and 6.5
OVC	Table 5.4	SALI	Table 6.6
OVC EP per ENNI	Table 5.5	EUSA (ENNI and UNI)	Table 6.7
OVC per UNI	Table 5.6	OVC/EVC	Table 6.8
UNI	Table 4.3	Transport (ENNI)	Table 6.9

that MEF specifies attributes and parameters and ASR forms are used to order services and facilities using Common Language codes that would represent required attributes and parameters.

Comparing these tables, one can find that attributes like CoS, bandwidth profile and service-level specifications (frame delay, frame delay variation, frame loss probability, and so forth) listed in tables in Chapters 4 and 5 have not been unambiguously mapped to Common Language codes in ASR forms. Similarly common language codes listed in ASR forms like TOS, LOS, P-Bit, BDW, NC, NCI, SPEC, SECNCI, and so forth have not been unequivocally mapped to attributes in MEF forms. This leads to confusion during implementation. This situation is quite understandable due to the fact that peering of CENs based on ENNI and using E-access service type is relatively a new field. It is expected that this situation will get rectified as additional implementation level guidelines are made available.

6.3 Linkage Between ASR Forms and IT Systems Related to OSS/BSS

As discussed earlier, access service request (ASR) is preceded by selling activities and leads to a long chain of activities that involves ordering design, order management, provisioning, installation coordination, preservice testing, service turn up, maintenance, and billing and ongoing coordination of testing and trouble resolution for all operator-provided facilities. Automation or mechanization of these and some additional activities is the function of large and complex IT systems called operations and business support systems (OSS/BSS). We will cover OSS/BSS systems in the next chapter.

6.4 Chapter Summary

Selling carrier Ethernet (CE) services to off-net customers over peering carrier Ethernet networks (CENs) require service providing operators to purchase E-access service from access providing operators. This purchase order, known as access service request (ASR), constitutes a business-to-business (B2B) transaction. To ensure that there is no confusion about what is being requested, standardization of these request forms is required. This chapter provided brief descriptions of various access service request forms available from the Ordering and Billing Forum (OBF) which manages these forms and is part of an industry organization called the Alliance for Telecommunications Industry Solutions (ATIS). The chapter then identified those forms that are specifically used for E-access service with the caveat that operators have some deviations in implementing these forms. Details of fields in those E-access service-specific forms were then provided in this chapter. While describing these fields, the chapter identified some Common Language codes needed in these ASR forms. These Common Language codes are maintained by an industry organization called iConnective. Because MEF specifies attributes and parameters and ASR forms use Common Language codes to order required attributes and parameters, this chapter compared MEF-specified attributes and parameters with ASR required Common Language codes and identified areas where further clarification in mapping of attributes and parameters to Common Language codes would help in the implementation of this new and fast emerging E-access service for peering CENs. Because the access service request (ASR) is preceded by selling activities and leads to a long chain of activities that includes design, ordering, order management, installation coordination, preservice testing, service turn-up, maintenance, and billing and ongoing coordination of testing and trouble resolution for all operator-provided facilities, automation, or mechanization of these activities is highly desirable. Large and complex IT systems called operations and business support systems (OSS/BSS) provide automation or mechanization of these and some more activities. This chapter provided a transition to OSS/BSS systems which is the topic of our Chapter 7.

7

OPERATIONS AND BUSINESS SUPPORT SYSTEMS

The function of good software is to make the complex appear to be simple.

Remarks by Grady Booch, Co-Developer of Unified Modeling Language

In the last chapter, we discussed that the access service request (ASR) is preceded by selling activities and succeeded by a long chain of activities that includes design, ordering, installation coordination, pre-service testing, service turn-up, maintenance and billing, and ongoing coordination of testing and trouble resolution for all operator-provided facilities. Automation or mechanization of these activities is highly desirable. Large and complex IT systems called operations and business support systems (OSS/BSS) provide automation or mechanization of these and some more activities. Because OSS/BSS systems include multitude of operations, a framework is essential to understand, architect, design, develop, integrate, and operate this complex group of IT systems. This framework is also helpful, like a "North Star," to guide service providers, operators, and vendors of these OSS/BSS systems to improve their current implementation. In this chapter, we will cover the framework of OSS/BSS systems and its evolution.

The chapter then covers additional functionalities that a well-designed, robust, secure, and redundant OSS/BSS system must support in view of the important toll services offered by service providers and operators. These additional functionalities are related to state machine, transactional integrity, security, and high availability. These are in addition to the aforementioned OSS/BSS functions related to verification, qualification, order-entry (OE), activation, billing, and service level agreement (SLA) management activities.

The chapter then dwells on efforts by service providers to reduce operational costs by integrating many features with OSS/BSS systems including support for self-care, help-desk outsourcing, network function virtualization (NFV), and software-defined networking (SDN).

Peering Carrier Ethernet Networks. http://dx.doi.org/10.1016/B978-0-12-805319-5.00007-1

From the descriptions in all the chapters so far, we know that data networks, leading up to peering CENs, move bits from source to destination, and the OSS/BSS systems provide automation or mechanization of operational and business processes. Data networks and OSS/BSS are very important and necessary foundations; however, the main purpose of all these foundations is to support applications or services that customers need, use, and pay for. This chapter provides a transition to customer applications which is covered in detail in the next chapter.

7.1 Evolution of OSS/BSS Framework

In the last chapter, we discussed that the ASR leads to a long chain of activities, many of which are automated by the operations and business support systems (OSS/BSS). As a matter of fact, many activities in OSS/BSS start even before ASR issuance. These preprovisioning activities are related to validating address where service is required, qualifying customers, showing available products and services to customers so they can choose what service they want and then performing credit checks, getting a signed contract or agreement from customers, and finally ordering the service. These activities are applicable to not just the E-access services but for all other types of products and services offered by a service provider or by an operator. Therefore, the role of OSS/BSS is much broader than just E-access services for peering CENs.

Fig. 7.1 shows a schematic of a typical OSS/BSS system. Using this Fig. 7.1, we will examine a typical sequence of transactions, just to illustrate the functions of an OSS/BSS system, for ordering a typical product, including but not limited to a CE product. The process starts with an inquiry for product or service from a prospective customer, which could originate from any one of the sales channels like a retail store or by calling toll free phone number or from a self-service web portal. All these channels access the operations and business support systems (OSS/BSS) for processing.

The first thing that the BSS system does is to prompt for address. After validating the address against Master Street Address Guide (MSAG) or a database based on GIS, the system then shows the listing of products and their bundles available at that location from Products & Pricing catalog. Once the customer selects a product, the system then prompts for additional information needed for credit rating verification and then calls the API of the credit rating agencies. This is needed by telecom service providers because they want to know what risk they are assuming by having that customer. This credit rating will

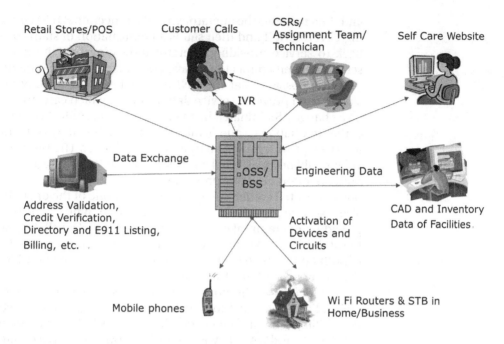

Retail Stores/POS   Customer Calls   CSRs/ Assignment Team/ Technician   Self Care Website

IVR

Data Exchange

OSS/ BSS

Engineering Data

Address Validation,
Credit Verification,
Directory and E911 Listing,
Billing, etc.

Activation of
Devices and
Circuits

CAD and Inventory
Data of Facilities

Mobile phones

Wi Fi Routers & STB in
Home/Business

Figure 7.1 Role of OSS/BSS.

determine how much deposit, if any, will be required from the customer and whether the customer will be billed before the month or after the month. Next, the BSS system generates an agreement or contract for customer to sign. Then, the BSS system generates an ASR and sends it to the OSS system. The OSS system on receiving an ASR accesses the engineering data to determine network facilities that are needed for making connections to complete the access circuit from customer location to their Central Office (CO) or head-end and from there to switch and, if needed, route data to the destination. A technician, if needed, is dispatched to complete all the connections and also to install a customer premises equipment (CPE). Next, a configuration or boot file is sent to CPE, and if a wireless connection has been ordered, then a profile file is sent to customer's mobile device to associate it with an eNodeB. After this, if needed, data is exchanged with third party vendors for directory listing, E911 services, Directory Assistance and so forth. And finally, a trigger is sent to the billing system to start the billing. This, in short, completes the activation.

In case of E-access services, when an ASR is received by an access provider, the access provider provisions a user network interface (UNI) and an external network–network interface (ENNI) and then provides the operator virtual connection (OVC) connecting UNI to ENNI and then sends a confirmation of status of ASR back to service providing operator indicating that the E-access service-related connections have been made. It should be noted that it is possible that the UNI and/or ENNI are already provisioned from previous service definitions, in which case, there will be a change ASR initiated instead of a new start ASR. As discussed earlier in Chapters 5 and 6, the MEF architecture does not require a service provider to be an operator or a carrier. The service provider could be an external entity such as a systems integrator, and both operators could be providing access service. However, in this book, as stated earlier, we will assume that one of the operators is also a service provider. With that assumption in place, the service providing operator takes similar steps using its OSS/BSS systems to establish an OVC in its CEN and then concatenates the OVCs of both operators to peer CENs needed for the E-line or E-LAN or E-Tree service ordered by a subscriber.

It is clear that the operations and business support systems (OSS/BSS) are needed for verification, qualification, OE, activation, billing, and SLA management of access in local access loop, MAN, RAN, and WAN. In view of the important toll services offered by service providers, a well-designed, robust, secure, and redundant OSS/BSS system is essential to do the following

1. Provide customer relationship management
2. Manage product and service catalog
3. Process access service request (ASR)
4. Manage flow-through provisioning
5. Manage customer care
6. Proactively monitor network elements, network, systems and applications
7. Intelligently analyze the collected data and provide fault and performance alarms
8. Take preventive measures
9. Ensure security
10. Manage SLA implementation, and
11. Manage billing

Obviously, OSS/BSS, shown as a box in Fig. 7.1, is a technologically complex system involving millions of lines of codes and interfaces with multitude of systems. Therefore, a framework is essential to understand, architect, design, develop, integrate, and operate this complex group of IT systems. This framework is also helpful, like a "North Star," to guide service providers, operators, and vendors of these OSS/BSS systems to improve their current implementation.

In 1996, the International Telecommunication Union–
Telecommunication Standardization Sector (ITU-T) released
recommendation M.3010,[41] introducing the concept of the
Telecommunication Management Network (TMN). The recommendation M.3010 was later revised and further expanded in
M.3013. This TMN framework was developed to facilitate service providers to manage their service delivery networks. It
consisted of different levels of abstraction namely, functional,
physical, informational, and logical. The logical level was further abstracted, as shown in Fig. 7.2, into four layers including
business management layer, service management layer, network management layer, and element management layer. The
physical level of abstraction included the network element layer
consisting of physical network elements which are activated,
monitored, and from which performance data is collected. In
1997, ITU-T published recommendation M.3400 extending the
TMN framework by introducing the functional level of abstraction based on the concept of fault, configuration, accounting, performance, and security (FCAPS). TMN framework was
an important start based on logical, physical, and functional
abstraction. It, however, did not include processes of telecom
service providers.

The development of the business process standardization
for telecom service providers started with the founding of the
TM Forum (formerly named TeleManagement Forum) in 1988.
Between 1995 and 1999, the TM Forum developed Telecom
Operations Map[42] (TOM). The focus of TOM was on the core
customer operational processes of fulfillment, assurance, and
billing (FAB). The goal of TOM was to create a framework of
business processes, including the definition of a common
enterprise independent terminology for service management.
It was also supposed to serve as a basis for discussing the scope
of information management necessary for the execution of the
processes. Fig. 7.2 shows a composite view of TMN and TOM
frameworks.

From Fig. 7.2, it is clear that both TMN and TOM needed
improvements. The TMN framework was discretized and not integrated. This did not allow systems belonging to different layers
to communicate for coordinating all the processes from strategy,
planning, and product creation to access provisioning, billing,
and customer care. TOM on the other hand had the limitation
that it only included FAB and did not include processes related to
operations support and readiness which is the "back-office" environment that enables support and automation for FAB.

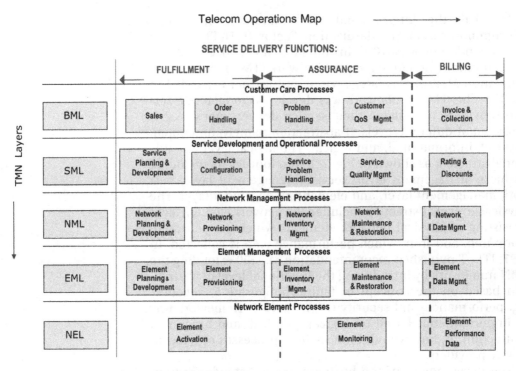

Figure 7.2 Combined TMN and TOM frameworks.

Fig. 7.3 below shows processes that are essential for an integrated approach. The infrastructure lifecycle management, product lifecycle management, supply chain lifecycle management, and

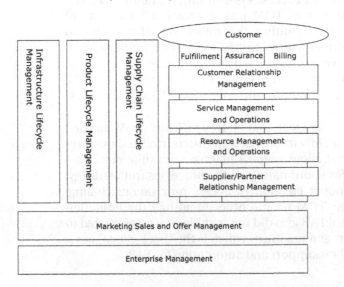

Figure 7.3 Integrated model for OSS/BSS.

enterprise management processes shown in Fig. 7.3 do not directly support the customer but are essential to the overall business.

There was another issue with TMN and TOM frameworks that required resolution. This related to distributed information or data sources. Many applications belonging to OSS/BSS systems had their own databases with fragmented data modeling for same object and this led to lot of translation of data back and forth. This situation is shown in Fig. 7.4.

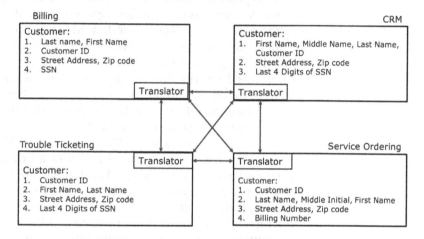

Figure 7.4 Example of fragmented data modeling.

To resolve this issue of fragmented data modeling, there was a need for shared data or information as shown in Fig. 7.5.

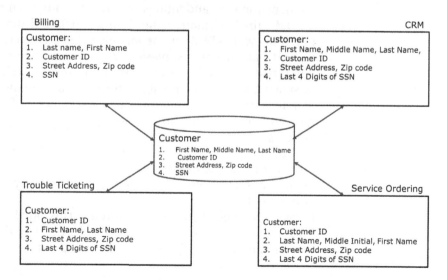

Figure 7.5 Shared information data model.

To address these issues, TMF started the New Generation Operations Systems and Software (NGOSS) specification. NGOSS, now renamed as Frameworx, is a comprehensive, integrated framework for developing, procuring, and deploying operations and business support systems (OSS/BSS)–related software. It is available as a toolkit of industry-agreed specifications and guidelines that cover key business and technical areas including:

- Business process framework (eTOM): An industry-agreed set of integrated business process descriptions, created with today's customer-centric market in mind, used for mapping and analyzing processes required for operation of the business.
- Shared information/data (SID) framework: Comprehensive, standardized information definitions acting as the common language for all data to be used in Frameworx-based applications. A common information language is the linchpin in creating easy-to-integrate software solutions.
- Telecom applications map (TAM) framework: TAM provides a framework for applications. It provides a bridge between eTOM and SID on one hand and real deployable and potentially procurable applications on the other hand. TAM does this by grouping process functions and information data into recognized OSS/BSS applications. TAM also provides a common language for communities who specify, procure, design, and sell operation and business support systems (OSS/BSS), so that they can understand each other's viewpoints
- Integration framework: The integration framework identifies the dependencies and unifies the business process framework (eTOM), the information framework (SID), and the application framework (TAM) in a service-oriented architecture (SOA) context that ensures seamless migration to a service-oriented enterprise. Integration framework also provides well-defined business and system language, interfaces, and architecture to facilitate system integrators with a clear direction for repeatable and cost-effective integration of multivendor, disparate systems.
- Technology neutral architecture (TNA): TNA provides key architectural guidelines and specifications to ensure high levels of transaction flow-through among diverse systems and components.
- Compliance and conformance criteria: This provides guidelines and tests to ensure that systems defined and developed using Frameworx specifications will interoperate.

- Lifecycle and methodology: This provides processes and artifacts that allow developers and integrators to use the toolset to develop Frameworx-based solutions using a standard approach.

The goal of Frameworx is the rapid development of flexible, low-cost-of-ownership solutions to meet the business needs of today's service providers and operators. Frameworx presents a number of benefits to various industry players, such as service providers, OSS/BSS software vendors, and systems integrators. It is important to understand that neither Frameworx as a whole nor its components, including eTOM, SID, TAM, and integration framework, are interfaces or data. They are not APIs. One cannot "write" some code using them. They are frameworks or "maps" only.

The Telecom Operation Map (TOM) was extended in 2001 and was first named eTOm and later was renamed as enhanced telecom operations map (eTOM); the change in name is very subtle and one needs to look carefully to notice the small difference! It was renamed again in 2013 as "Business Process Framework (eTOM)." There have been many versions of eTOM. Version 14 was published in May 2014. The "Business Process Framework (eTOM)" describes and analyzes different levels of enterprise processes according to their significance and priority for the business. This framework is defined as generically as possible so that it remains organization, technology, and service independent. For service providers, the "Business Process Framework (eTOM)" serves as the blueprint for process direction. It also provides a neutral reference point for internal process re-engineering needs, partnerships, alliances, and general working agreements with other companies. For suppliers, the "Business Process Framework (eTOM)" outlines potential boundaries of software components that should align with their customers' needs, clearly highlighting the required functions, inputs, and outputs that must be supported by their products. At the overall conceptual level, the "Business Process Framework (eTOM)" can be viewed as having the following three major process areas:

- Strategy, infrastructure, and product (SIP)—covering planning and lifecycle management
- Operations—covering the core of day-to-day operational management
- Enterprise management—covering corporate or business support management

This is shown in Fig. 7.6 below. These three major business process areas cover seven end-to-end vertical process groupings required to support customers and manage the business. Among these vertical groupings, the focus of eTOM is on the core customer operational processes of fulfillment, assurance, and billing (FAB). Operations support and readiness is the "back-office" environment that enables support and automation for FAB. The SIP processes do not directly support the customer, and they include the strategy and commitment and the two lifecycle process groupings. Enterprise management also does not support customers directly but is essential for running business and includes planning, risk management, financial management, HR management, knowledge management, and effectiveness management.

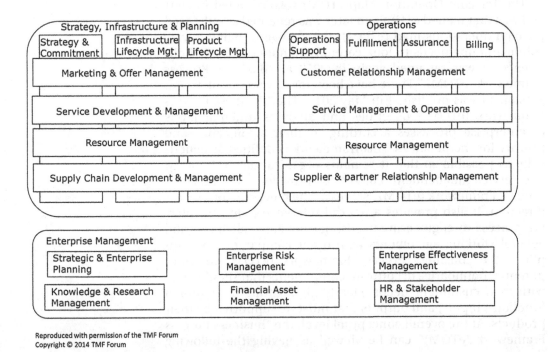

Figure 7.6 The business process (eTOM) framework.
TM Forum http://www.tmforum.org.

SID model of the Frameworx (NGOSS) is shown in Fig. 7.7 below. SID is based on decomposition of relevant information into five primary domains market/sales, product, customer, service, resource, supplier/partner, and enterprise.

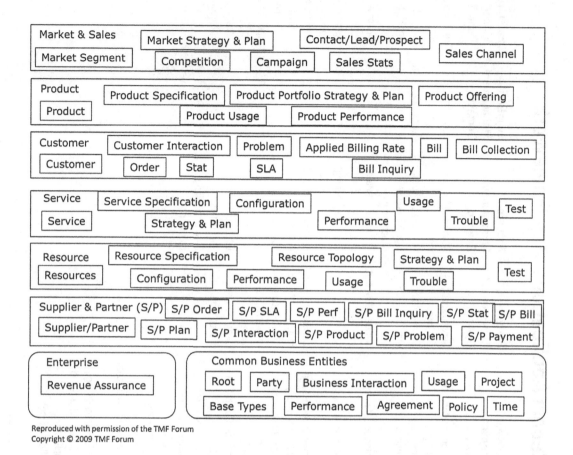

Figure 7.7 Shared information/data framework.
TM Forum http://www.tmforum.org.

Goal of SID is to provide a logical view of things of interest (entities) to an enterprise, along with corresponding relationships (associations), facts (attributes), and behavior (operations).

Telecom Applications Map (TAM) is another framework which is part of Frameworx (NGOSS). It is shown below in Table 7.1. TAM provides a frame of reference for applications. It provides a bridge between eTOM and SID on one hand and real deployable and potentially procurable applications on the other hand. It does this by grouping process functions and information data into recognized OSS/BSS applications.

TAM also provides a common language for communities who specify, procure, design, and sell operation and business support systems (OSS/BSS), so that they can understand each other's viewpoints.

Table 7.1 Telecom Applications Map Framework

Market/Sales Domain	Product Management Domain	Customer Management Domain	Service Management Domain	Resource Management Domain	Supplier Management Domain	Enterprise Management Domain
Campaign Mgt.	Product strategy mgt.	Customer information mgt.	Service specification mgt.	Resource lifecycle mgt.	Partner mgt.	Revenue assurance mgt.
Sales aids	Product/service catalog mgt.	Translational document production	Service inquiry mgt.	Resource process mgt. (workflow integration)	Supply chain mgt.	HR mgt. applications
Compensation & results	Product lifecycle mgt.	Customer order mgt.	Service order mgt.	Resource inventory mgt.		Financial mgt. applications
Mass market sales mgt.	Product performance mgt.	Customer self mgt.	SLA mgt.	Resource order mgt.		Asset mgt. applications
Corporate sales mgt.		Customer contact mgt., retention & loyalty	Service problem mgt.	Resource assurance mgt.		Security Mgt. applications
Sales portals		Customer service rep toolbox	Service quality monitoring & impact analysis	Voucher mgt.		Knowledge mgt. applications
		Customer CoS & SLA mgt.	Service performance mgt.	Billing data mediation applications		Fraud mgt. applications
		Customer service/ account problem resolution		Real-time billing mediation		
		Receivable mgt.				
		Billing account mgt.				
		Billing inquiry & dispute mgt.				
		Bill format rendering				
		Product/service rating Application				
		Collection mgt.				
		Bill calculation				
		Online charging applications				

The integration framework is shown in Fig. 7.8. The integration framework identifies the dependencies and unifies the "Business Process Framework (eTOM)," the information/data framework (SID), and the application framework (TAM) in a SOA context that ensures seamless migration to a service-oriented enterprise. Integration framework provides well-defined business and system language, interfaces, and architecture to facilitate system integrators with a clear direction for repeatable and cost-effective integration of multivendor, disparate systems.

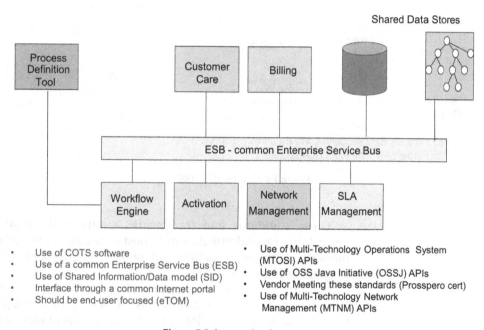

Figure 7.8 Integration framework.

In the past, the integration was based on proprietary systems like RMI, CORBA, or DCOM. With the evolution of Web services, these proprietary systems have been replaced by open Web services which are based on SOA. Also, the adoption of J2EE standard in middleware or workflow engine allowed for system-level abstraction including containers for servlets and beans, services such as JDBC, JMS, JNDI, JMail, JTA, JLog, and support for multimedia service. SOA also includes standards like SOAP, WSDL, and Universal description, discovery, and integration (UDDI). The SOA is shown in Fig. 7.9 below.

<u>Definition of SOA</u>

According to W3C, SOA is a set of components which can be invoked and whose interface descriptions can be published and discovered. Where the terms are defined as below

- "A set" means a collection of services that can work together on a common infrastructure
- "Components" means modular implementation
- "invoked" means a client can call the services
- "interface description" means definition of the interface
- "published and discovered" means consumer of services is separated from provider of services such that the consumer can locate the services registered in a registry

<u>Example of SOA based on Web Services</u>

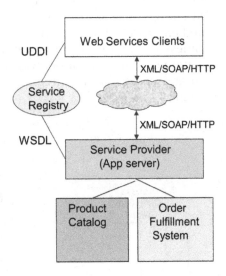

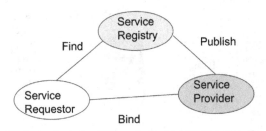

- SOA also includes E2E security, Load Balancing, Template Driven Workflow/BPM and Data Normalization

Figure 7.9 SOA for OSS/BSS.

These standards and frameworks contributed significantly toward the use and integration of best of breed, commercial off-the-shelf software in building operations support and business support systems (OSS/BSS).

It is interesting to note that some telecom providers are also monetizing their OSS/BSS systems by selling or renting their OSS/BSS systems to smaller operators in various modes like perpetual license or as cloud services including IaaS, PaaS, or SaaS. They are also offering additional BPO services such as bill printing and mailing, customer support and trouble ticket management, designing, developing, and managing self-service Web portals, using location awareness for targeted advertisements and so forth.

Now that we have examined the functions of OSS/BSS and various frameworks that provide guidelines to architect, design, develop, integrate, and operate OSS/BSS; in the next section, we will discuss some important features of the operations and business support systems (OSS/BSS). These features are essential for the functioning of this business critical system consisting of many complex applications.

7.2 State Machine, Transactional Integrity, Security and High Availability of OSS/BSS

7.2.1 State Machine

An important capability of this OSS/BSS system is to handle move, add, change, and disconnect of the services popularly known in short as MACDs. In addition, the customer premises equipment (CPE) that supports multiple services like data, voice, and video services may go through preprovisioned and provisioned states and after the subscriber is active, each of the services may go through suspend or abuse states. For example, if a subscriber is found to abuse data service then that service can be put under suspension due to abuse, but the service provider is legally obliged to keep limited voice service so that the subscriber can make emergency calls to E911. As a result of all these permutations and combinations, a subscriber and subscriber's services and the associated customer premises equipment (CPE) go through many states. Kangovi[14] developed a state machine called service-linked multi-state system (SLIMS) to handle all these multiple states. This state machine plays an important role in the OSS/BSS system.

7.2.2 Transactional Integrity

Because OSS/BSS performs multiple functions, manages events and resources, and enables all parts of the system to work coherently, and it is critical to the business, therefore, it must be able to scale with high throughput and must be reliable. The necessary condition for the system to be reliable is an assured end-to-end communication with all applications in the system working properly and not ending up in an error condition. OSS/BSS can be thought of as made of north bound and south bound systems. The north bound systems mainly consist of customer relationship management, order management system, and so forth and south bound systems include provisioning systems, billing systems, device activation systems, service assurance systems, and so forth. Transactions usually start from the north bound system and are received by a workflow engine which then converts the transactions to a workflow, that is, a series of tasks in a given order for the south bound systems. The workflow engine only maintains the state of the customer during the transaction execution. Once the transaction is completed, the state is not maintained in the workflow engine. The end result of this is that the subscriber and his/her services/devices are provisioned or deleted or changed as the case may be, provided the transaction reached its destination without impairment of its function, content, or meaning. The integrity of the transaction also ensures that all the data stores reflect correct state of the data.

Different types of transactions traversing the network and the definition of a well-behaved system include three categories of transactions namely, synchronous, asynchronous, and bulk.

Synchronous transactions are processed in a serial fashion. In this type of transactions, the source system sends the transaction to middle ware on a thread and waits for response. The middle ware then spawns tasks for other systems and on completion of all the tasks and sends the response back to the source system. The billing systems operate in this manner. In the event that there is an error in processing, queuing will occur in the billing system until the error is cleared. Excessive queuing will degrade the performance of the billing systems for all transactions.

Asynchronous transactions are processed by queuing the request. The source system queues the transaction on the queue of the middle-ware, which then processes the queue by FIFO or LIFO basis. After completing the request, the middle-ware sends the response to a queue of the source system. Occasionally, in case of asynchronous transactions, one part of a sequence of the transactions is executed asynchronously, and other parts of the sequence are processed synchronously. This is because of dependencies on target systems to complete the sequence synchronously on the same thread. Once target systems have completed the balance of transactions and send the response to a queue of the source system then asynchronous part of the sequence is completed. The process of provisioning a device is a collection of synchronous threads executed asynchronously as described briefly in the following steps:

- The first sequence registers the transaction with the provisioning system. At this point, all that is known about the customer is the services requested. Remaining provisioning requests related to this customer will be handled after the handshake between billing and provisioning has completed.
- The second sequence occurs when the device is plugged in at the customer premise and is confirmed by the OSS/BSS that the device is legitimate and granted access to the network. Other device-related requests are executed on the same processing thread only after the device is confirmed in the device provisioning system.
- The final sequence takes place when the billing is updated and confirmed, and cycle of billing begins.

Some business scenarios require the ability to support loads of massive numbers of transactions at a time. Bulk operations are supported via specific API interfaces. Usually, bulk transactions are used for maintenance activities like market area split or merge, zip code changes, area code changes, or preprovisioning of devices for market trials of some new service that has not yet been integrated with billing.

A canonical form of the provisioning system, described earlier and consisting of north bound, middleware, and south bound systems, is shown below in Fig. 7.10. It shows the request transaction starting from a north bound system to the middleware, which in turn starts a task for a south bound system. On receiving the response from the south bound system, the middleware sends a response back to the north bound system. This completes the flow of transaction.

According to ATIS (Alliance for Telecommunications Industry Solution) Transaction Integrity is defined as "the degree to which a transaction flowing through a network reaches its intended destination without impairment of its function, content or meaning".

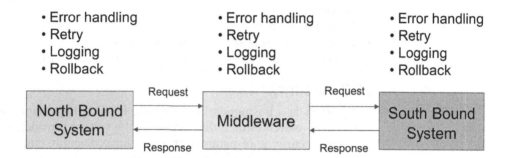

Figure 7.10 A typical ideal system.

In an "ideal" system, error handling, retry, logging, and rollback capabilities are implemented in every subsystem as shown in Fig. 7.10.

According to ATIS[39] (Alliance for Telecommunications Industry Solution), transaction integrity is defined as "the degree to which a transaction flowing through a network reaches its intended destination without impairment of its function, content or meaning". Transaction integrity is a property of a total collection of data. It cannot be maintained simply by using reliable primitives for reading and writing single units because the relations between the units are also important. Therefore, to protect transaction integrity, one must architect with data in mind and consider the following:

- How often is data queried? Where?
- How often is data updated? By whom?
- How much data is manipulated? Is there a cacheable subset?

Transaction integrity is just one of several QoS (quality of service) elements of OSS/BSS, other elements include security and process orchestration. Transaction integrity monitoring is needed for operational governance to meet that QoS of OSS/BSS. The approach to ensuring transaction integrity must be relatively simple and must have modest demands on the underlying system. The suggested technique is to ensure that data integrity will not be lost as a result of communication failures or application errors. The systems should be designed to implement consistent, atomic transactions coupled with error handling, logging, retry, and rollback strategies to achieve transaction integrity. The errors could be intra-system errors or inter-system errors. The strategies to ensure transaction integrity include:

- Get an audit of the entire system
- Examine business impact of transactions for manual corrective actions
- Keep data store in clean state
- Determine correct sequence of operations
- Generate reports of errors and manually correct them

7.2.3 Security

OSS/BSS is a business critical system. It holds important customer data as well as data that is vital to the service provider, and therefore security is of paramount importance. Security considerations must prevent distributed denial of service (DDoS), man in the middle attack, IP spoofing, unauthorized port scanning, packet sniffing, and unauthorized configuration management type of attacks. This is achieved by zoning of IT systems as shown in Fig. 7.11 below.

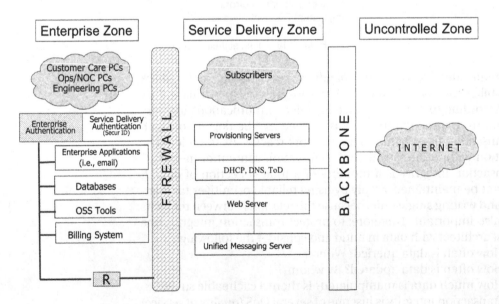

Figure 7.11 Implementation of zoning for security of IT systems.

In addition to this implementation of zoning of IT systems, the physical security of data centers is very important as well. It is shown in Fig. 7.12. Compliance with various security and privacy-related standards and specifications are required. Some of these include Sarbanes-Oxley, ISO 27001, PCI DSS, SA770 (SOCI), FISMA A&As, DIACAP MAC III Sensitive ATO, and so forth. Obviously, security is a large and very specialized field, and detailed description is beyond the scope of this book.

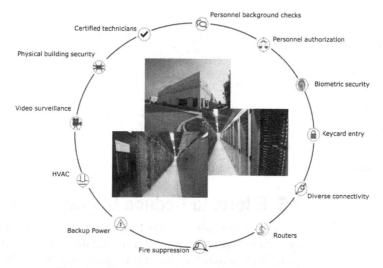

Figure 7.12 Security considerations in data center design.

7.2.4 High Availability of OSS/BSS

Because many services including voice, data, and video that are critical to millions of customers are being offered by service providers, it is becoming critical to ensure high availability of not only the networks but also that of OSS/BSS systems to meet SLAs. This requires implementing load balancing, redundancy, and disaster recovery. Redundancy and Disaster recovery are implemented by periodic back-ups and also by having secondary data centers. Use of load balancers and multithreading provides additional high availability and redundancy functions. This is shown in Fig. 7.13 below.

As multiple services including Voice, Data and Video are being offered by Service Providers, it is becoming critical to ensure High Availability of the Network and Systems in order to meet SLAs. This requires implementing Redundancy and Disaster Recovery.

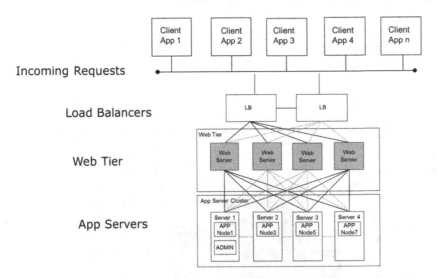

Figure 7.13 High availability of OSS/BSS using load balancers and multithreading.

7.3 Efforts to Reduce Operational Costs

Service providers today face formidable challenges. As customers add new services, the volume of data traffic traversing the service provider's network continues to grow exponentially. This is driving the cost of networks up with demand. At the same time, revenue per user is remaining flat or increasing at a much smaller rate thus impacting profitability. This situation is depicted in Fig. 7.14 below.

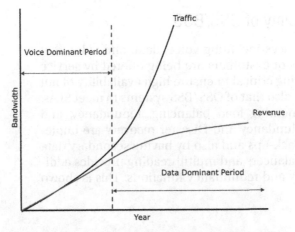

Figure 7.14 Bandwidth growth versus revenue for operators.

This is putting pressure on the service providers to reduce operating costs. Service providers are using various means to achieve this goal. One of them is to out-source help-desk function. Fig. 7.15 shows the schematic of the out-sourced help-desk function. This requires integration of the IVR and call center system with the OSS/BSS systems so that remotely located call centers can validate customers, enter trouble tickets, resolve issues, and close trouble tickets.

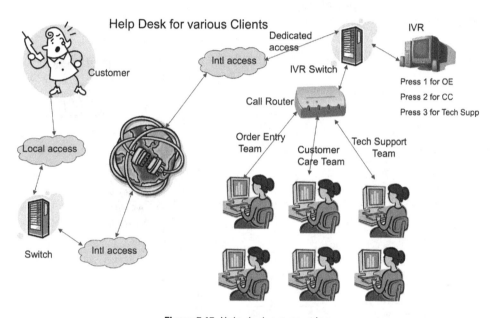

Figure 7.15 Help-desk out-sourcing.

Another technique that service providers are using is to encourage self-provisioning by customers. As shown in Table 7.2, the cost savings from self-provisioning is considerable. Self-service moves contacts from high-cost call center channels to lower cost IVR and web channels.

Table 7.2 Rationale for Self-care Portal of Order Entry

Service Channel	Cost Factor	Baseline Cost
CAE (agent)	1.0	$6.00
IVR	0.08	$0.48
Channel portal	0.25	$1.25
E-mail	0.40	$2.40
Chat/IM	1.0	$6.00

Focus on self-service via IVR and Web allows for the highest and best use of capital and operating resources. Additional channels can be used as alternatives to call centers or self-care portals but have not proved as effective as the IVR and self-service Web portals in delivering savings. For example, the cost for e-mail and Web chat interactions is close to that for live telephone interactions. Fig. 7.16 shows a typical implementation of self-service portal.

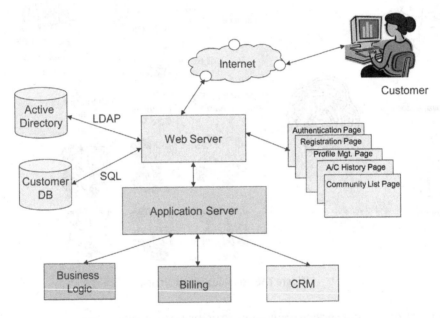

Figure 7.16 A typical self-care portal logical architecture.

Analytics is another area that service providers are using to improve efficiency and quality of decision making to make business competitive and reducing costs. It is shown in Fig. 7.17. It involves extracting data from various OSS/BSS systems, staging the data, loading the transformed data in data marts and enterprise data warehouses, and then integrating these with analytic applications to create dashboards and score cards. The service providers are also using historical data for data mining to determine patterns using artificial intelligence applications and using predictive analytics for predicting future trends.

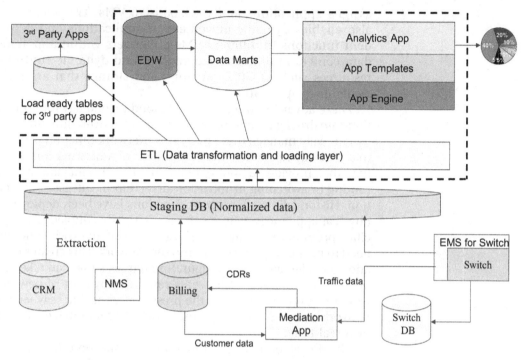

Figure 7.17 Architecture for analytics from OSS/BSS systems.

There are two emerging areas that offer huge potential for further reducing operating and capital costs and also for implementing on-demand services. These include NFV and SDN. These are currently getting lot of attention because NFV and SDN will allow operators to use cheaper generic hardware and allow customers to not only order the services but also configure them as well in near real-time. We will cover these topics and their impact on OSS/BSS in the next section.

7.4 Impact of NFV and SDN on OSS/BSS

The popularity of cloud technology for applications is motivating efforts to virtualize many network functions and to utilize customer driven on-demand processes for ordering and configuring network services. Cloud technology is based on the fact that it provides the ability to dynamically manage workloads of diverse applications. Cloud technology leverages real-time instantiation of virtual machines (VMs) on commercial hardware where appropriate and then dynamically assigns applications and workloads to VMs. This technology also enables dynamic movement of applications and dependent functions to different VMs on servers within and across data centers in different geographies and dynamic control of resources such as CPU, memory, and storage that are made available to applications.

These advances and benefits of cloud technology for applications are driving efforts to virtualize many network functions and dynamically manage these functions. Network function is a term that typically refers to some component of a network infrastructure that provides a well-defined functional behavior, such as routing or switching or intrusion detection or intrusion prevention. Historically, such network functions have been deployed as physical appliances, where software is tightly coupled with specific, proprietary hardware. These physical network functions need to be manually installed into the network. This creates operational challenges with configuration requests that can take days or weeks to handle and prevents rapid deployment of new network functions especially in the present circumstances where the network team is constantly being bombarded for rapid changes in near real-time.

The focus of network function virtualization (NFV) is to transform network elements into software applications running on generic hardware. Before getting into detail, let us examine the definitions of the terms NFV and virtual network function (VNF). Many use these terms interchangeably, which can be a source of confusion. NFV specifications, set forth by European Telecommunications Standards Institute[43] (ETSI), makes it clear that the acronyms have related but distinct meanings. NFV typically refers to the overarching principle or concept of running software-defined network functions, independent of any specific hardware platform. NFV also refers to a formal network virtualization initiative led by some of the world's biggest telecommunication network operators. In conjunction with ETSI, these companies aim to create and standardize an overarching, comprehensive NFV framework, a high-level illustration of which is shown in Fig. 7.18. It shows a physical network, a cloud that has VNFs and virtual networks for subscribers that tunnel through the same physical network.

VNF, on the other hand, refers to the implementation of a network function using software that is decoupled from the underlying hardware. This can lead to more agile networks with significant Opex and Capex savings. A VNF is a software that represents a network function such as firewall or IDS/

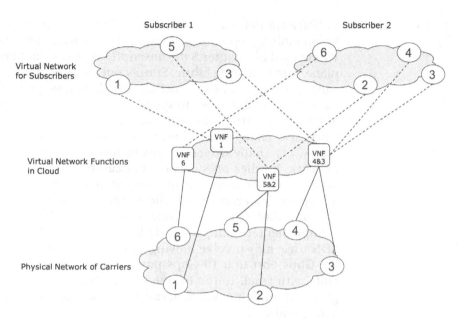

Figure 7.18 Network function virtualization.

IPS, or load balancer or a switch or a router. Just as a system administrators or a subscriber can set up a VM by pointing and clicking, similarly, they can turn up a firewall or IDS/IPS using a VNF. VNF uses best practices as base policies and configurations for different network elements. If a subscriber is creating a specific tunnel or virtual network through the infrastructure, as shown in Fig. 7.18, they can add a firewall or IDS/IPS or a switch to just that tunnel. VNFs run on high-performance platforms (currently x86 platforms dominate this category) in a cloud, and it enables users to turn up functions on selected tunnels in the network. The goal is to allow subscribers to create a service profile for a VM and leverage power of x86 platforms to build an abstraction on top of the network and then build services on that specific logical environment. Once in place, VNF saves a lot of time on manual provisioning and training. VNF also reduces the need to overprovision because rather than buying big firewall or IDS/IPS boxes that can handle a whole network, the customer can buy functions for the specific tunnels that need them. This reduces initial Capex, however, the real advantage stems from cost reduction in operational area. VNF can be thought of as a parallel to cloud, with a few boxes running a lot of virtual servers and a point and click provisioning system.

Software-defined networking (SDN) makes the network programmable by separating the control plane from the data plane. We covered in Chapter 5 the descriptions of control plane, data plane, and switching fabric. Simply put, control plane tells network device, primarily a switch, which frames go where and data plane sends frames to specific destinations. Although NFV and VNF add virtual tunnels and functions to the physical network, SDN changes the physical network and therefore is really a new, externally driven means to provision and manage the network. It relies on switches that can be programmed by an SDN controller using an industry standard control protocol. SDN provides a way to consolidate the control plane into a single controller with appropriate redundancy rather than having it implemented independently in each switch. A use case for SDN's use may involve moving a circuit with large traffic from a 1-Gbps port to a 10-Gbps port, or aggregation of many circuits with small traffic to one 1-Gbps port. SDN controller is generally implemented in data centers and not on x86 servers in VNF clouds.

The implementation of NFV, VNF, and SDN will have profound impact on OSS/BSS system. In the near term, OSS/BSS will have to support both legacy physical network functions (PNFs) and the emerging VNFs as shown in Fig. 7.19.

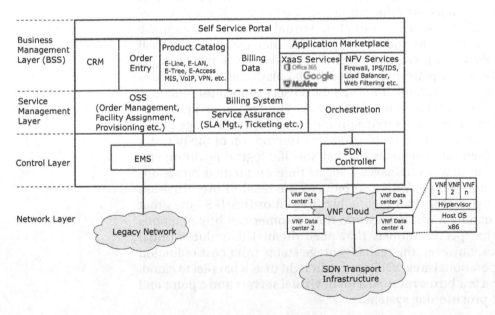

Figure 7.19 SDN and OSS/BSS for delivering NFV.

This situation is further complicated by the fact that many service providers have OSS/BSS systems that are legacy systems where billing systems are not able to bill for services provided on peering CENs and NFV services. It is not simple to replace these well-entrenched legacy billing systems because, for example, billing account numbers are being used in these legacy OSS/BSS systems as primary key for provisioning systems. Resolution of these issues is prerequisite to integrating OSS/BSS with VNF cloud and SDN to provide NFV over peering CENs.

7.5 Customer Applications Are the Business of Operations/Business Support Systems

From the descriptions in all the chapters so far, we know that data networks, leading up to peering CENs, move bits from source to destination, and the OSS/BSS systems provide automation or mechanization of operational and business processes. Data networks and OSS/BSS are very important and necessary foundations; however, the main purpose of all these foundations is to support applications or services that customers need, use, and pay for. Examples of such services or applications include mobile backhaul, multi-site connectivity, connectivity to clouds, high-speed internet connections, VoIP, VPN, connections for cyber-physical systems, and so forth. In the next chapter, we will cover some of these applications and their dependencies on MEF-defined E-line, E-LAN, and E-tree services requiring E-access services based on peering CENs.

7.6 Chapter Summary

This chapter described the architectural framework and functions of operations and business support systems (OSS/BSS) proposed by industry organizations including ITU-T and TM Forum. These business critical systems are a large and complex group of IT applications and provide automation or mechanization of multitude of activities before and after an access service request (ASR). In Chapter 6, it was covered that to provide services to off-net subscribers, E-access type of service is needed on peering CENs based on ENNI and to order that service an access service request (ASR) needs to be submitted to an access providing operator. This ASR results in a long chain of activities that includes design, ordering, installation coordination, preservice testing, service turn-up, maintenance and billing, and ongoing coordination of testing and trouble resolution for all operator-provided facilities. Operations and business support systems (OSS/BSS) provide highly desirable automation or mechanization of these and some more activities. Because OSS/BSS systems include multitude of operations and is a large and complex group of IT applications, a framework is essential to understand, architect, design, develop, integrate, and operate this system. This framework is also helpful, like a "North Star," to guide service providers, operators, and vendors of these OSS/BSS systems to improve their current implementation.

The chapter then covered additional functionalities a well-designed, robust, secure, and redundant OSS/BSS system must support, in view of the important toll services offered by service providers. These additional functionalities are related to state machine, transactional integrity, security, and high availability. These are in addition to the functions related to verification, qualification, OE, activation, billing, and SLA management activities.

The chapter then described efforts by service providers to reduce operational costs by integrating many features with OSS/BSS systems including support for self-care, help-desk outsourcing, network function virtualization (NFV), and software-defined networking (SDN).

From the descriptions in all the chapters so far, we know that data networks, leading up to peering CENs, move bits from source to destination, and the OSS/BSS systems provide automation or mechanization of operational and business processes. Data networks and OSS/BSS are very important and necessary foundations; however, the main purpose of all these foundations is to support applications or services that customers need, use, and pay for. This chapter provided a transition to customer applications which is covered in detail in the next chapter.

8

APPLICATIONS OF PEERING CARRIER ETHERNET NETWORKS

"Information is giving out, communications is getting through."

Remarks by Sydney J. Harris, an American journalist

Descriptions in all previous chapters show that Carrier Ethernet networks (CENs) and their peering as well as operations and business support systems (OSS/BSS) are very important and necessary foundations. We also learned that the primary purpose of all these foundations is to support various applications or services that customers need, use, and pay for. In order to understand these applications, we first need to understand major categories of customers and then dwell on the applications used by them. This chapter, therefore, starts with the taxonomy of customers and their applications.

The chapter then presents application-specific performance requirements or objectives compiled by Metro Ethernet Forum (MEF) from variety of sources in public domain. These application-specific performance objectives (APOs) are then mapped to MEF-defined standard CoS (Class of Service) performance objectives (CPOs) and performance tiers (PTs). This mapping is crucial to standardizing Carrier Ethernet (CE) services particularly in peering CENs. This mapping is also described in this chapter.

The chapter then covers the applicable CE services including Ethernet-access (E-Access) service for peering CENs to meet the network functionality needed by customer applications. Examples of network functionality include IP backhaul, mobile backhaul, streaming and switched video transport, site-to-site connectivity, connection for cloud computing services, and network connectivity for emerging applications such as Internet of things (IoT), cyber-physical systems (CPSs), and virtual reality (VR).

The chapter next dwells, briefly, on a process to convert information about customer applications and topology into a design for a CE service based on CENs and peering CENs. Finally, the chapter transitions to next steps needed in peering of CENs to accommodate emerging trends. This is covered in some detail in the final chapter of this book.

Peering Carrier Ethernet Networks. http://dx.doi.org/10.1016/B978-0-12-805319-5.00008-3

8.1 Taxonomy of Customers and Applications

The main purpose of CENs and peering CENs, access service request (ASR), and large and complex OSS/BSS systems is to support applications or services that customers need, use, and pay for. In order to understand these applications, we first need to understand major categories of customers and then dwell on the applications used by them. The taxonomy of the customers and their applications is shown in Fig. 8.1. This figure shows customer categories as consumers (this includes fixed and mobile consumers), enterprise, and mobile operators. This is slightly different from that given in MEF 23.1[36] which identifies three main categories of customers, namely, consumers, business, and mobile. Fig. 8.1 also shows categories of applications depending on if they are layer 1, layer 2, or layer 3 applications, and then Fig. 8.1 lists some examples of applications under each category. This taxonomy shown in Fig. 8.1 then depicts few examples of network functions in access loop and also in backhaul. Based on Section 1.4 of Chapter 1, we know that CENs and peering CENs are generally deployed in backhaul.

Fig. 8.1 is not a complete taxonomy of all possibilities, but it serves the purpose to illustrate the hierarchy that will be useful in understanding subsequent sections. Table 8.1 lists the applications and their customers referenced from MEF 23.1.[36] It is important to note

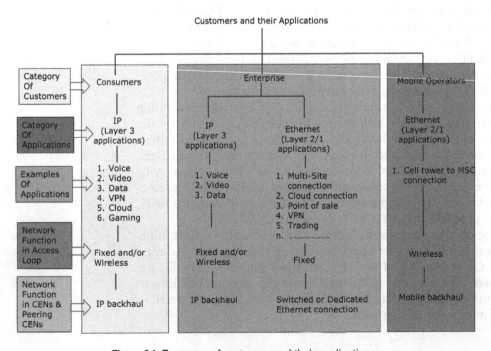

Figure 8.1 Taxonomy of customers and their applications.

Table 8.1 Mapping of Applications to Customers

Application	Consumer	Enterprise	Mobile
VoIP data	X	X	X
Interactive video (videoconferencing)	X	X	?
Web browsing	X	X	X
IPTV data plane	X	X	?
IPTV control plane	X	X	?
Streaming media	X	X	X
Interactive gaming	X	-	X
Best effort	X	X	X
Circuit emulation	-	X	X
Telepresence	-	X	-
Remote surgery (video)	-	X	-
Remote surgery (control)	-	X	-
Telehealth (Hi-res image file transfer)	-	X	-
Email	X	X	X
Broadcast engineering (Pro-video over IP)	-	X	-
CCTV	X	X	X
Financial/trading	-	X	-
Database	-	X	-
Real-time fax over IP	X	X	-
Store-and-forward fax over IP	X	X	-
SANs (synchronous replication)	-	X	-
SANs (asynchronous replication)	-	X	-
Wide-area file services (WAFS)	-	X	-
Network-attached storage	X	X	-
Text terminals (telnet, ssh)	-	X	-
Graphics terminals (thin clients)	-	X	-
Point of sale transactions	-	X	-
E-commerce (secure transactions)	X	X	X
Mobile backhaul system requirements	-	-	X

CCTV, closed-circuit TV; *IPTV,* IP television; *SAN,* storage area network; *VoIP,* voice over IP (Internet Protocol).
Reproduced with permission from Metro Ethernet Forum.

that the information in the MEF specifications are either normative or informative. The information given in Table 8.1 is informative.

After identifying the applications, MEF compiled APOs from variety of sources in public domain including standard-based references, industry-based best practices, vendor-specific information, and product-specific information. The resulting APOs from MEF 23.1[36] are also informative and shown in Table 8.2.

The performance parameters shown in Table 8.2 including frame delay (FD), FD variation (FDR), mean frame delay (MFD), frame loss ratio (FLR), and interframe delay variation (IFDV) have been covered in Section 4.4 of Chapter 4, and frame delay has been explained in Fig. 4.14 in Chapter 4. In Table 8.2, Pd is frame delay performance percentile, Pr is specific percentile of the frame delay performance used in FDR, Pv is IFDV performance percentile, T is a time interval that serves as a parameter for service-level specification, and Δt is a time interval much smaller than T.

Comparing Tables 8.1 and 8.2, one can notice that not all applications in Table 8.1 have been included in Table 8.2 because MEF could not find clear guidance for performance requirements for those applications. Second, one will also notice that there are many applications in Table 8.2 for which some performance objectives are marked as "not specified" because of the unavailability of data. Nevertheless, Tables 8.1 and 8.2 represent important information based on lot of research on part of MEF.

As we discussed in Chapters 4 and 5, MEF was formed to standardized carrier-grade services that are scalable, reliable, and support quality of service. Therefore, the next logical step involves mapping these APOs to standard CPOs and PTs defined by MEF and described in Chapters 4 and 5, so that customers could compare offerings from various providers and also to facilitate service-level agreement (SLAs) between service providers and customers. This mapping is critical for peering CENs. This is the topic of the next section.

Table 8.2 Application-Specific Performance Objectives

Application	FD	MFD	FLR	FDR	IFDV
VoIP data	125 ms prefer 375 ms limit Pd=0.999	100 ms prefer 350 ms limit	3e-2	50 ms Pr=0.999	40 ms Pv=0.999
Interactive video (videoconferencing)	125 ms prefer 375 ms limit Pd=0.999	100 ms prefer 350 ms limit	1e-2	50 ms Pr=0.999	40 ms Pv=0.999

Table 8.2 Application-Specific Performance Objectives—continued

Application	FD	MFD	FLR	FDR	IFDV
VoIP and video conference signaling	Not specified	250 ms prefer	1e-3	Not specified	Not specified
IPTV data plane	125 ms Pd = 0.999	100 ms	1e-3	50 ms Pr = 0.999	40 ms Pv = 0.999
IPTV control plane	Not specified	75 ms	1e-3	Not specified	Not specified
Streaming media	Not specified	Not specified	1e-2	2 s	1.5 s Pv = 0.99
Interactive gaming	50 ms	40 ms	1e-3	10 ms	8 ms
Circuit emulation	25 ms Pd = 0.999999	20 ms	1e-6	15 ms Pr = 0.999	10 ms Pv = 0.999, Δt = 900 s, T = 3600 s
Telepresence, includes remote surgery (video)	120 ms Pd = 0.999	110 ms	2.5e-4	40 ms Pr = 0.999	10 ms
Financial/trading	Unknown	2 ms	1e-5	Unknown	Unknown
CCTV	150 ms (MPEG-4) 200 ms (MJPEG) Pd = 0.999	Not specified	1e-2	50 ms Pr = 0.999	Not specified
Database (hot standby)	5 ms	Not specified	1e-5	Unknown	Unknown
Database (WAN replication)	50 ms	Not specified	1e-5	Unknown	Unknown
Database (client/server)	Not specified	1 s	1e-3	Not specified	Not specified
T-38 fax	400 ms Pd = 0.999	350 ms	3e-2	50 ms Pr = 0.999	40 ms Pv = 0.999
SANs (synchronous replication)	5 ms	3.75 ms	1e-4	1.25 ms	1 ms
SANs (asynchronous replication)	40 ms	30 ms	1e-4	10 ms	8 ms
Network-attached storage	Not specified	1 s	1e-3	Not specified	Not specified
Text terminals (telnet, ssh)	Not specified	200 ms	1e-3	Not specified	Not specified

Continued

Table 8.2 Application-Specific Performance Objectives—continued

Application	FD	MFD	FLR	FDR	IFDV
Point of sale transactions	2 s	1 s	1e-3	Not specified	Not specified
Best effort includes:	Not specified	Not specified	Not specified	Not specified	Not specified
Email					
Store/forward fax					
Wide-area file services (WAFS)					
Web browsing					
File transfer (including hi-res image file transfer)					
E-commerce					
Mobile backhaul (high CoS)	10 ms	7 ms	1e-4	5 ms	3 ms
Mobile backhaul (medium CoS)	20 ms	13 ms	1e-4	10 ms	8 ms
Mobile backhaul (low CoS)	37 ms	28 ms	1e-3	Not specified	Not specified

CCTV, closed-circuit TV; *IPTV*, IP television; *MJPEG*, Motion JPEG; *MPEG*, Moving Picture Experts Group; *SAN*, storage area network; *VoIP*, voice over IP (Internet Protocol); *WAN*, wide area network.
Reproduced with permission from Metro Ethernet Forum.

8.2 Mapping of Application Performance Objectives to Standard MEF CoS Performance Objectives

In Chapter 5, Section 5.6, we covered that the color of the frame can be marked by drop eligibility identifier (DEI) bit in virtual LAN (VLAN) tags by setting it to 0 for green and 1 for yellow frames. We also covered in Section 5.6 that priority code point (PCP) values can also be used to mark color of the frame. MEF 23.1 recommended normative three CoS labels are shown in Table 8.3 along with the corresponding PCP values in VLAN tag for green and yellow color frames.

MEF 23.1 also defined four normative PTs as shown in Table 8.4. In fact, MEF 23.2, which is in a draft state currently, is proposing to add a fifth PT. This normative PT definition uses

Table 8.3 CoS Implementation Agreement Model Proposed by MEF

CoS Label	PCP Value in VLAN Tag for Green Color	PCP Value in VLAN Tag for Yellow Color
High (H)	5	Not specified
Medium (M)	3	2
Low (L)	1	0

Reproduced with permission from Metro Ethernet Forum.

Table 8.4 Performance Tier (PT) Implementation Agreement

PT	Distance	Propagation Delay	Remarks
City (PT 0.3)	<75 km	0.6 ms	Proposed in MEF 23.2 (draft)
Metro (PT 1)	<250 km	2 ms	Normative in MEF 23.1
Regional (PT 2)	<1200 km	8 ms	Normative in MEF 23.1
Continental (PT 3)	<7000 km	44 ms	Normative in MEF 23.1
Global (PT 4)	<27500 km	172 ms	Normative in MEF 23.1

Reproduced with permission from Metro Ethernet Forum.

distance as the primary means of categorizing PTs and for deriving minimum delays. The distance stated for each PT can be considered as approximate distance limit. This PT reference model was chosen to allow for sufficient granularity and also to cover. CENs and their peering in metropolitan area network (MAN), regional area network (RAN), wide area network (WAN), and even global networks. The propagation delay (PD) shown in Table 8.4 was derived by assuming;

PD = 0.005 ms/km × 1.25 × distance in kilometers,

where 0.005 is the propagation time in milliseconds per kilometer, and 1.25 is circuit path–to–airline distance ratio. The PD was rounded off to the nearest integer in Table 8.4. Distance is difficult to ascertain in real networks as path (i.e., circuit) distance is unknown or may vary due to routing or other path changes (e.g., dynamic control protocols). In CENs and peering CENs, there may be additional delays due to, for example, switch hops, buffering, shaping, serialization for low speed links, etc. These are not included in the aforementioned equation to calculate PD.

Assumptions made in MEF 23.1 and MEF 23.2 in arriving at this PT Implementation Agreement (IA) shown in Table 8.4 include the following:
1. PT distances represent the path a frame would traverse and thus drive associated PD minimums for FD/MFD/FDR.
2. Though number of switch hops generally increases with longer distance PTs, hops will not be quantified.
3. For simplicity, PT CPOs are expressed as constants based on the maximum distance for the PT rather than formulas with distance variables.
4. PTs are derived with certain distance and application assignments.
5. PTs can be arbitrarily assigned to given services by operators based on factors in or outside the scope of this IA.
6. All links, including access links, will have a link speed of at least 10 Mbps, with the notion that a given service may utilize a "higher" PT for slower links based on operator discretion.

As covered in Section 5.6 of Chapter 5, after defining CoS Labels and PT, MEF 23.1 specified the normative CPOs for each combination of CoS Label and PT as shown in Table 5.13 and reproduced for the sake of continuity in this chapter in Table 8.5 with the addition of PT 0.3 from MEF 23.2 for point-to-point case.

It is important to note that MEF 23.1 has defined these normative CPOs only for point-to-point cases. Normative CPOs for multipoint are proposed in the MEF 23.2 and are shown in Table 8.6.

CPOs for rooted multipoint cases are slated to be developed by MEF later.

Now that APOs were compiled (Table 8.2) from information available in public domain and standard CPOs were defined (Tables 8.5 and 8.6), the next step for MEF was to map the applications to CPOs. For this MEF developed, a CPO compliance tool using Microsoft Excel spreadsheet to test candidate CPO values against the APOs shown in Table 8.2 and applied a set of statistical

Table 8.5 MEF Specification of Point-to-Point CoS Performance Objectives for Performance Tiers (PTs)

CoS Label PT	CoS Label H					CoS Label M					CoS Label L				
	PT 0.3	PT 1	PT 2	PT 3	PT 4	PT 0.3	PT 1	PT 2	PT 3	PT 4	PT 0.3	PT 1	PT 2	PT 3	PT 4
FD (ms)	≤3	≤10	≤25	≤77	≤230	≤6	≤20	≤75	≤115	≤250	≤11	≤37	≤125	≤230	≤390
MFD (ms)	≤2	≤7	≤18	≤70	≤200	≤4	≤13	≤30	≤80	≤220	≤9	≤28	≤50	≤125	≤240
IFDV (ms)	≤1	≤3	≤8	≤10	≤32	≤2.5 or n/s	≤8 or n/s	≤40 or n/s	≤40 or n/s	≤40 or n/s	n/s	n/s	n/s	n/s	n/s
FDR (ms)	≤1.25	≤5	≤10	≤12	≤40	≤3 or n/s	≤10 or n/s	≤50 or n/s	≤50 or n/s	≤50 or n/s	n/s	n/s	n/s	n/s	n/s
FLR (%)	≤0.001	≤0.01	≤0.01	≤0.025	≤0.05	≤0.001	≤0.01	≤0.01	≤0.025	≤0.05	≤0.1	n/s	n/s	n/s	n/s

n/s, not specified.
Reproduced with permission from Metro Ethernet Forum.

Table 8.6 MEF Specification of Multipoint CoS Performance Objectives for Performance Tiers (PTs)

CoS Label PT	CoS Label H					CoS Label M					CoS label L				
	PT 0.3	PT 1	PT 2	PT 3	PT 4	PT 0.3	PT 1	PT 2	PT 3	PT 4	PT 0.3	PT 1	PT 2	PT 3	PT 4
FD (ms)	≤3	≤10	≤25	≤77	≤230	≤6	≤20	≤75	≤115	≤250	≤11	≤37	≤125	≤230	≤390
MFD (ms)	≤2	≤9	≤20	≤72	≤202	≤5	≤15	≤32	≤82	≤222	≤10	≤30	≤52	≤127	≤242
IFDV (ms)	≤1	≤3	≤8	≤10	≤32	≤2.5 or n/s	≤8 or n/s	≤40 or n/s	≤40 or n/s	≤40 or n/s	n/s	n/s	n/s	n/s	n/s
FDR (ms)	≤1.25	≤5	≤10	≤12	≤40	≤3 or n/s	≤10 or n/s	≤50 or n/s	≤50 or n/s	≤50 or n/s	n/s	n/s	n/s	n/s	n/s
FLR (%)	≤0.001	≤0.01	≤0.01	≤0.025	≤0.05	≤0.001	≤0.01	≤0.01	≤0.025	≤0.05	≤0.1	≤0.1	≤0.1	≤0.1	≤0.1

n/s, not specified.
Reproduced with permission from Metro Ethernet Forum.

and other constraints to the candidate CPO values to make sure that they maintain the correct relationships to each other across CoS labels, across PTs, and between the CPOs within a single CoS label/performance tier. The candidate CPO values were modified as necessary to meet the constraints while still satisfying the APOs. MEF 23.1[36] lists the following constraints on CPOs to avoid a statistical contradiction:

1. $FDR > FD - MFD$
2. $MFD < FD$
3. $IFDV < FDR$
4. $FD - MFD \gg 0.5 \, FDR$ (0.5 represents a symmetric distribution, this constraint ensures a gradual slope to right.)

   MEF 23.1 also describes two additional constraints to ensure consistency between the values for FD and FDR and the estimated maximum PD associated with each performance tier. When the percentile parameter $Pd = Pr$, then the minimum delay (MinD) associated with a given CoS label/performance tier can be calculated as $MinD = FD - FDR$. This value MinD should not be less than PD. MinD should also not be significantly higher than PD. This constraint is satisfied by;

5. $FD - FDR \geq PD$

   The second constraint is satisfied if the CPO values meet either of two tests. The first test scales PD by a ratio and then compares it to MinD. The second test, which prevents the constraint from becoming too severe for very low PDs, adds a fixed offset to PD before comparing it to MinD. The second test is expressed as

6. $(FD - FDR \leq PD \times 1.5)$ OR $(FD - FDR \leq PD + 20 \, ms)$

   Finally, for PT constraints, it was assumed that CPOs should never improve as tier number increases.

   Each CPO value is compared to the corresponding APO value. If the CPO value is less stringent than the APO value, it is considered not compliant, otherwise, the CPO value is considered compliant. MEF also defined two types of compliance levels namely, loose and tight. If the APO value is within a configurable range of the CPO value, it is considered tight compliance; otherwise, it is loose compliance. As an example, if an APO for MFD is 50% higher or less stringent than the corresponding CPO, it is considered in loose compliance. An unspecified or unknown application objective also results in loose compliance.

   The resulting mapping of application to MEF-based standardized CoS labels and PTs is shown in Table 8.7.

Table 8.7 Mapping of Applications to CoS IA Labels and Performance Tiers

Category of Application	CoS Label / Performance Tier	H 0.3	H 1	H 2	H 3	H 4	M 0.3	M 1	M 2	M 3	M 4	L 0.3	L 1	L 2	L 3	L 4
Real time	VoIP data	X	X	X	X											
Interactive	VoIP and videoconferencing signaling			X	X	X								X	X	X
Real time	Video conferencing data						X	X	X	X	X					
Near real time	IPTV data plane						X	X	X	X						
Interactive	IPTV control plane						X	X	X	X						
Streaming	Streaming media	X					X				X					
Low delay	Interactive gaming						X	X								
Very low delay	SANs (synchronous replication)	X											X			X
Low delay	SANs (asynchronous replication)							X								
Best effort	Network-attached storage							X					X	X	X	X
Best effort	Text and graphics terminals							X					X	X	X	X
Near real time	T.38 fax over IP	X						X		X	X					
Very low Delay	Database hot standby								X							
Low delay	Database WAN replication							X								
Low delay	Database client/server													X	X	
Low delay	Financial/trading	X						X		X	X					
Near real time	CCTV	X						X		X	X					
Real time	Telepresence	X		X	X			X								
Real time	Store-and-forward fax over IP	X		X							X					
Real time	Circuit emulation	X						X								
Very low delay	Mobile backhaul—high (H)												X			
Very low delay	Mobile backhaul—medium (M)															
Low delay	Mobile backhaul—low (L)											X				

CCTV, closed-circuit TV; IPTV, IP television; VoIP, voice over IP (Internet Protocol); SAN, storage area network.
Reproduced with permission from Metro Ethernet Forum.

This mapping in conjunction with CPOs given in Tables 8.5 and 8.6 will provide CPOs for each application. For example, voice over IP (Internet Protocol) (VoIP) data which is a real-time application are mapping to CoS label H, and PT1 in Table 8.7 has CPO values of $FD \le 10\,ms$, $MFD \le 7\,ms$, $IFDV \le 3\,ms$, $FDR \le 5\,ms$, and $FLR \le 0.01\%$ from Table 8.5.

Going back to the taxonomy diagram shown in Fig. 8.1, now that the applications are mapped to standard MEF-defined CPOs, next step is to map network-specific functionality required by applications to CE-based standard services including Ethernet Private Line (EPL), Ethernet Virtual Private Line (EVPL), Ethernet Private Local Area Network (EP-LAN), Ethernet Virtual Private LAN (EVP-LAN), Ethernet Private Tree (EP-Tree), and Ethernet Virtual Private Tree (EVP-Tree) and including E-Access services for off-net sites for peering CENs. This is the topic of next section.

8.3 Mapping Application-Specific Network Functionality to Carrier Ethernet Services

8.3.1 IP Backhaul

Consumers of fixed and mobile services mostly use their services for Internet access, VoIP, legacy telephony, access to TV and media streams, and virtual private network (VPN) services.

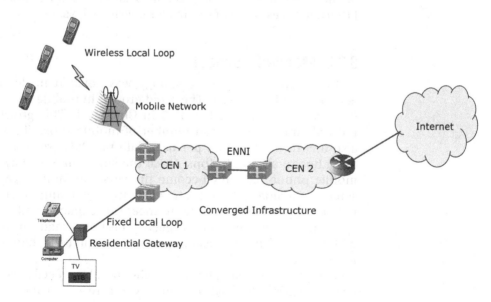

Figure 8.2 Carrier Ethernet services for IP backhaul.

Increasingly, all these services are built on the IP suite of protocols, and service provisioning to an end user increasingly becomes equivalent of enabling the flow of IP traffic to and from the subscribers, as shown in Fig. 8.2. Much of the telecommunications operator's efforts today are centered on creating seamless services working over both the fixed and the mobile networks defined by the term Fixed Mobile Convergence (FMC). One important element of the operator's FMC strategy is to use a common transport infrastructure of packet-optical equipment for all IP services, irrespective of whether access is fixed or mobile. This simplifies the implementation of common higher layer service entities and reduces both Capex and Opex. The infrastructure delivering the operator's services, as described in Chapters 1 and 2 and shown in Fig. 2.7, is typically divided into an access network, an aggregation network, and a core network. The role of the aggregation network is to transport IP traffic from a large number of access nodes to much fewer core nodes. Traditionally, this task has been performed by synchronous digital hierarchy (SDH)/ synchronous optical network (SONET) links and wavelength-division multiplexing (WDM) wavelengths, but a more capacity-efficient aggregation network is accomplished when using CE services to aggregate multiple traffic streams. CE leverages the inherent capabilities for statistical multiplexing of much of the data traffic and fills the available bandwidth pipes more efficiently. In case of off-net customers, CE services also allow the mechanism of Ethernet-access type of service for peering CENs.

8.3.2 Mobile Backhaul

Evolution of mobile technology was covered in detail in Section 1.4 of Chapter 1. The rapid growth in mobile data traffic was also shown in Fig. 1.18 in Chapter 1. This growth is not only due to growth in number of mobile subscribers but also due to the fact that per-user data rate has grown from a few Kbps to about 25 Mbps. Once the single-line display of a mobile phone has now become the versatile multimegapixel screen of a smartphone (note: currently, the common screen is 1242×2208 pixel). In order to meet this capacity and coverage requirements, new radio spectrum is being made available with smaller cell sizes resulting in more radio cells than just a few years ago.

This growth in data rates and higher number of cells place an ever increasing demand for capacity and reach on the mobile backhaul networks. A primary challenge for mobile operators is to

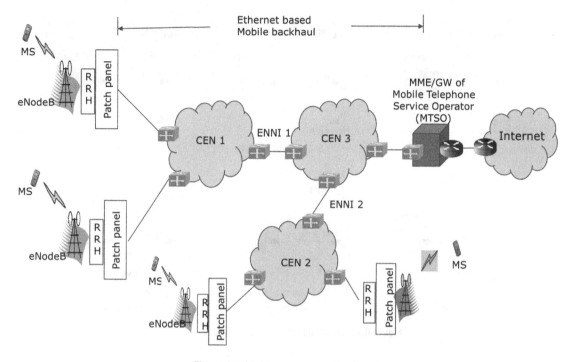

Figure 8.3 Mobile access and backhaul.

dimension mobile backhaul networks to cope with increasing traffic demand, avoiding making it a bottleneck for mobile services, and avoiding end-user frustration over long response times and unpredictable performance. With the introduction of IP-based 3G and 4G/ Long Term Evolution (LTE) mobile networks, the demand for a CE-based aggregation network and peering CENs have become very attractive to mobile operators. LTE, in turn, introduces new requirements on how to synchronize and maintain the network. Moreover, mobile operators require transport services with more stringent SLAs. This in turn has opened up new and important business opportunity to operators sitting on large assets of fiber. A typical schematic for mobile backhaul requiring peering CENs is shown in Fig. 8.3. It shows cell sites marked as eNodeB connected to access aggregation CENs which are then peered to transport CEN connecting to the mobile operator's switching center. The peering of CENs is often needed because cell towers are generally located in one operator's territory and mobile switching center is located in another operator's territory.

Currently, in vast majority of cases, CENs and peering CENs for mobile backhaul are based on Ethernet Virtual Connections (EVCs) with single CoS because it is easier to implement. This single CoS

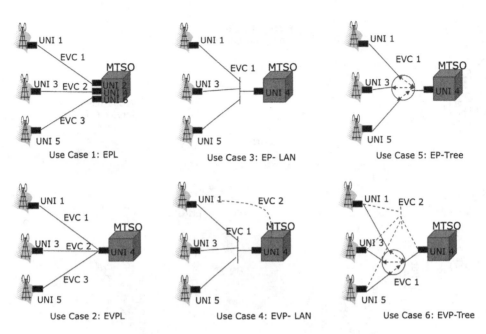

Figure 8.4 Various use cases in mobile backhaul.

is set at high to meet performance objectives for voice. However, smartphones handle all types of multimedia applications, and in fact, the bandwidth used by voice is much lower than other applications which have lower performance objectives. In view of this, a multi-CoS mobile backhaul is more desirable. This may increase the operational complexity but will reduce the Capex and Opex for both mobile operators and the backhaul or CEN operators.

Depending on a mobile operator's requirements, different CE service types can be deployed in mobile backhaul. MEF 22[36] describes many such use cases. These use cases are shown in Fig. 8.4. For the sake of simplicity, only mobile operator's (customer's) perspective is shown in the figure. Use case 1 in Fig. 8.4 shows the EPL service–based mobile backhaul. In this case, all untagged, priority-tagged, and tagged service frames are mapped to one EVC at the User Network Interface (UNI). The EPL service might be preferred in cases where there is a desire for a 1-to-1 port level correspondence between cell tower and mobile switching center (MSC). Port-based EPL services with dedicated UNI ports at MSC for every cell tower base station is not a scalable model. Use case 2 in Fig. 8.4 shows the EVPL service–based mobile backhaul. This VLAN-based service is used to access multiple cell sites with service multiplexing at the UNI at the Mobile Telephone Switching Office (MTSO) site. This allows efficient use of the MSC UNI. Use case 3 in Fig. 8.4 shows the EP-LAN service–based mobile back-haul. This allows multiple MSC sites or deployments with intercell tower communication. In this configuration, all sites appear to be on the same LAN. A key advantage of this approach is that if the mobile operator has outsourced its backhaul network to a service provider, for example, transport network carrier, the mobile oper-ator can configure CE-VLANs at the MSC UNIs and the cell tower UNIs without any need to coordinate with the service provider. Use case 4 in Fig. 8.4 shows the EVP-LAN service–based mobile backhaul. This allows different CE-VLAN ID sets to be mapped to the different EVCs at the UNI with service multiplexing. Use case 5 in Fig. 8.4 shows the EP-Tree service–based mobile back-haul. Traditionally, in mobile backhaul, the UNIs at cell tower sites only need to exchange service frames with the UNI at the MSC and not with other cell tower UNIs. This behavior is possible in an EP-Tree service. This is similar to use case 1 (EPL-based mobile backhaul) but requires only one rooted multipoint EVC. Use case 6 in Fig. 8.4 shows the EVP-Tree service–based mobile backhaul. This allows mobile operators not only to keep the root–leaf rela-tionship between MSC and cell tower sites, but also to have ser-vice multiplexing at one or more of the interconnected UNIs. It is important to remember that all the use cases in Fig. 8.4 show subscriber's point of view, and subscriber in this case is a mobile operator. That is the reason, Fig. 8.4 does not show any CENs or peering CENs; these are relevant in CEN operator's point of view only. It is clearly evident from the previous discussions that mobile backhaul design has to be carefully evaluated to examine which configuration of CE will best meet the requirements of a mobile operator.

8.3.3 Streaming and Switched Video Transport

Evolution of hybrid fiber coaxial (HFC) network, covered in detail in Section 1.3 of Chapter 1, shows that a modern MSO offers a wide range of services over the transport infrastructure once built for TV distribution. Services encompass both enter-tainment and communications for residential users as well as connectivity and value-added services to enterprises. An exam-ple of the HFC network was given in Fig. 1.12 in Section 1.3 of Chapter 1. A major trend in HFC networks is the rapid growth of traffic per end user and in the overall network. The emergence of streaming HD and 3D video services has made IP traffic grow some 30%–50% per year. To meet these demands, MSOs have gone through a conversion from analog to digital distribution technologies and packet-optical networks. Packet-optical net-works based on CENs have been the industry consensus answer

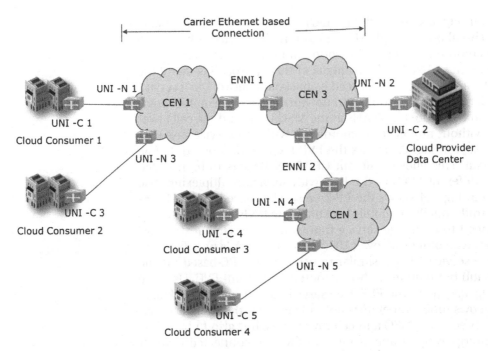

Figure 8.5 Carrier Ethernet services for cloud computing services.

to meet future needs, resulting from the convergence of legacy and next-generation services. Using CENs and peering CENs for integration of various functionalities into a common transport network equipment enables cost efficient capacity increases and also allows service differentiation capabilities.

8.3.4 Cloud Applications

The need to access data from anywhere on any device is rapidly making clouds the epicenter of consumers and businesses IT infrastructure today. Cloud could be private or public cloud. Public clouds like Amazon Web Services are based on shared physical hardware which is owned and operated by third-party providers, meaning that there are no hardware or maintenance costs incurred by cloud consumers. Private clouds, on the other hand, use bespoke infrastructure purely dedicated to a cloud consumer, hosted either on-site or at a service provider data center. The private cloud delivers all the agility, scalability, and efficiency of the public cloud but in addition provides greater levels of control and security, making it ideal for larger businesses and organizations. Cloud computing is pivotal to business agility and, of course, depends on data centers and

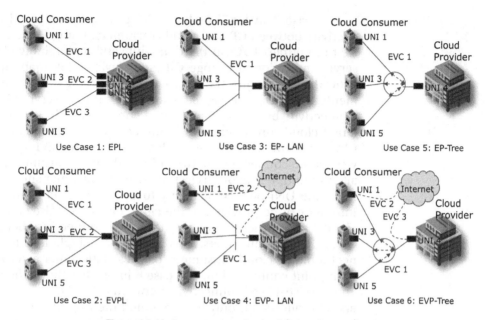

Figure 8.6 Various use cases in cloud computing application.

connectivity to data centers and between data centers. This connectivity is predominantly provided by Internet; however, larger enterprises are hesitant to move mission-critical applications to a cloud connected via the Internet due to concerns with security, network performance and strict data, regulation, and governance obligations. CE is ideal to address these concerns and more, whether it is about moving an application or a set of users between sites or invoking a disaster recovery plan, high-capacity flexible connectivity is crucial to businesses that run multitenant data center estates or that are moving toward cloud computing for their own business use. A typical schematic diagram of Ethernet services for cloud computing services is shown in Fig. 8.5 with peering CENs.

MEF 47[36] describes various use cases of the CE configurations for cloud computing application. These are shown in Fig. 8.6. For the sake of simplicity, Fig. 8.6 only shows cloud provider (CP)'s (customer's) perspective. Use case 1 in Fig. 8.6 shows the EPL service–based connectivity between CP and cloud consumers. This EPL service can provide elastic service behavior to change committed information rate (CIR) and flexible CoS mapping to allow the CP to adjust the service to meet varying traffic requirements. The EPL service can also be used for data center interconnect to provide dedicated bandwidth between CP data center sites.

Use case 2 in Fig. 8.6 shows the EVPL service–based connectivity between CP and cloud consumers. Using EVPL allows CP to multiplex EVCs at the UNI while still maintaining elastic service behavior to change CIR and flexible CoS mapping to allow the CP to adjust the service to meet varying traffic requirements. Use case 3 in Fig. 8.6 shows the EP-LAN service–based connectivity between CP and cloud consumers. This allows different cloud consumer sites to connect to each other as well as CPs' data centers. Use case 4 in Fig. 8.6 shows the EVP-LAN service–based connectivity between CP and cloud consumers. This allows multiplexing of services at a cloud consumer site so that the same UNI port connectivity to both CPs' data center and Internet can be provided to the cloud consumer. Use case 5 in Fig. 8.6 shows the EP-Tree service–based connectivity between CP and cloud consumers. This allows for a single EVC to connect multiple cloud consumers to a CPs' data center in a rooted multipoint configuration. Use case 6 in Fig. 8.6 shows the EVP-Tree service–based connectivity between CP and cloud consumers. This allows not only for the rooted multipoint connectivity between CP and cloud consumer sites but also for multiplexing of services at one or more cloud consumer sites, for example, to connect to the Internet.

It is important to remember that all the use cases in Fig. 8.6 show subscribers' point of view; subscriber in this case is a CP. That is the reason, Fig. 8.6 does not show any CENs or peering CENs; these are relevant in CEN operators' point of view only. It is clearly evident from the previous discussions that network design has to be carefully evaluated to examine which configuration of CE will meet the requirements of a CP best.

8.4 Cyber-Physical Systems and Other Emerging Applications

Internet started with Web 1.0 which refers to the first stage in the World Wide Web, which was entirely made up of web pages connected by hyperlinks. It then evolved into Web 2.0 which describes World Wide Web sites that emphasize user-generated content, usability, and interoperability. The Web 2.0 is now gradually evolving into IoT which refers to the ever-growing network of physical objects that feature an IP address for Internet connectivity and the communication that occurs between these objects and other Internet-enabled devices and systems. Just as this transformation is taking place, now CPSs are expanding IoT by integrations of computation, networking, and physical processes. Here

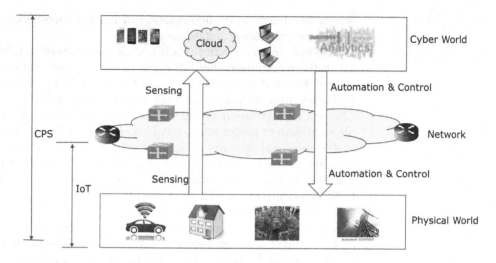

Figure 8.7 Schematic of a typical cyber-physical system.

embedded computers and networks monitor and control the physical processes, with feedback loops between physical processes and computations. CPSs are similar to the IoT sharing the same basic architecture; nevertheless, CPS presents a higher combination and coordination between physical and computational elements. CPS first evolved from deeply embedded systems which then led to IoT and now to CPS. It is a precursor to smart-connected planet.

A schematic diagram of the CPS is shown in Fig. 8.7. CPS is electromechanical devices augmented with communication capabilities and connectivity. It is divided into a cyber world and a physical world, and both are connected by a system of networks allowing sensor signal to be communicated from physical world to cyber world for analysis and then for handling of control and automation of the physical world. Today, a precursor generation of CPSs can be found in areas as diverse as aerospace, automotive, chemical processes, civil infrastructure, energy, healthcare, manufacturing, transportation, entertainment, and consumer appliances. CPS involves transdisciplinary approaches, merging theory of cybernetics, mechatronic design, and process science. The US National Science Foundation[44] has identified CPSs as a key area of research due to the potential of CPSs in several applications including, intervention (e.g., collision avoidance), precision (e.g., robotic surgery and nano-level manufacturing), operation in dangerous or inaccessible environments (e.g., search and rescue, firefighting, and deep-sea exploration), coordination (e.g., air traffic control, war fighting), efficiency (e.g., zero-net energy buildings), and augmentation of human capabilities (e.g., healthcare monitoring and delivery).

The link between cyber world and physical world in CPS has led to the recognition that advances in science and engineering are needed to dramatically increase the adaptability, autonomy, efficiency, functionality, reliability, safety, and usability of CPSs. CPS will also broaden mobile CPSs as well in cases where the physical system under study has inherent mobility. Examples of mobile physical systems include mobile robotics and electronics transported by humans or animals. The rise in popularity of smartphones has also increased interest in the area of mobile CPSs. Smartphones make ideal mobile CPSs interfaces because smartphones offer the following:

- significant computational resources, such as processing capability, local storage, multiple sensory input/output devices, such as touch screens, cameras, GPS chips, speakers, microphone, light sensors, proximity sensors;
- multiple communication mechanisms, such as Wi-Fi, 3G, EDGE, Bluetooth for interconnecting devices to either the Internet, or other devices;
- high-level programming languages that enable rapid development of mobile CPS node software;
- readily available application distribution mechanisms, such as the Android Market and Apple App Store;
- end-user maintenance and upkeep, including frequent recharging of the battery; and
- for tasks that require more resources than are locally available, one common mechanism for rapid implementation of smartphone-based mobile CPS nodes utilizes the network connectivity to link the mobile system with either a server or a cloud environment, enabling complex processing tasks that are impossible under local resource constraints.

Examples of mobile CPSs include applications to track and analyze CO_2 emissions, detect traffic accidents, insurance telematics, provide situational awareness services to first responders, measure traffic, and monitor cardiac patient.

The design of CPS and the implementation of their applications need to rely on IoT-enabled architectures, protocols and application programming interfaces (APIs) that facilitate collecting, managing, and processing large data sets, and support complex processes to manage and control such systems at different scales, from local to global. The large-scale nature of IoT-based CPS can be effectively and efficiently supported and assisted by cloud computing infrastructures and platforms, which can provide flexible computational power, resource virtualization, and high-capacity storage for data streams and can ensure safety, security, and privacy. The integration of networked devices, people, and physical systems is creating such a tantalizing vision of future possibilities that CPS is expected to become a vibrant part of the digital business landscape. There are other applications such as VR that will also emerge in next few years that will require higher bandwidths and near–real-time quality of service. Undoubtedly, CENs and peering CENs will have a vital role to play in CPS as well as other emerging applications.

Now that we covered taxonomy of customers and their applications and the mapping of these applications to MEF-defined standard performance objectives (CPOs) and CE services requiring peering CENs, in the next section, we will cover briefly the process of converting an application and topology information into a design for CE service.

8.5 Process for Converting Application and Topology Information into a Design

Description in this section leverages all the learnings from all the chapters in this book particularly, Chapters 4–8 and multitude of standards and specifications related to CEN, peering CENs, ASR, OSS/BSS, and customer-specific applications. In view of this, the process of design encompasses areas that are quite diverse and vast requiring large number of people with specialization in different fields. The design process is always a collaborative effort between various teams. The process starts with an examination of customer applications and topology of sites from connectivity perspective. From application, one can determine PT and CPOs as described in this chapter. From site topology one can determine if there is any off-net location; if yes, peering of CENs will be needed. Next, in discussion with the customer, the service provider determines bandwidth, CIR, and type of CE service including, EPL, EVPL, EP-LAN, EVP-LAN, EP-Tree, or EVP-Tree CE that will do the job. In case peering CENs are needed, then Ethernet-access service will be required, and as part of the design, service provider has to determine if Access E-Line or Access E-LAN type of Ethernet-access service is needed and then determine if the Ethernet-access service is going to be port based or VLAN based. For VLAN-based CE services, it is important to examine if UNI-based, EVC-based, or CoS-based bandwidth profile is needed. It should be noted that MEF 6.2 requires use of CoS-based bandwidth profile only. It is important to use S-tags in External Network Network Interface (ENNI) since it is defined as an S-tagged interface and to use Provider Backbone Bridge Traffic Engineering (PBB-TE)-based connection–oriented approach in the MAN/RAN/WAN. It is interesting to note that currently far more carriers use MPLS/MPLS-TP

(Multiprotocol Label Switching - Transport Profile). It may change in future with the wider acceptance of PBB-TE due to reasons that we discussed in Chapter 5. It is also important to base the design, as much as possible, on "switch many route once" approach to get improved performance.

Next section provides a transition to steps that are needed to further enhance and accelerate the implementation of peering CENs.

8.6 Next Steps

The description in this chapter shows that some applications, especially CPS and VR, are emerging. These applications as well as the emerging network function virtualization (NFV) and software-defined networking (SDN) will impact networks in a profound way. In addition, the APOs of these new applications need to be defined and mapped to MEF standard CPOs. It was also mentioned in Section 8.1 that there are gaps in applications to CPO mapping, and those need to be filled. Coupling of CPS and cloud computing based on NFV and SDN utilizing generic hardware and open-source VNFs puts extra importance on reliability, safety, and rigorous testing. In the next chapter, we will discuss these topics and next steps in furthering implementation of peering CENs to meet the changing landscape of customer applications and networks.

8.7 Chapter Summary

This chapter described the taxonomy of customers and their applications. This taxonomy provided a structure and important insight in understanding applications that customers need, use, and pay for. Customer applications are the primary purpose of CENs and peering CENs as well as OSS/BSS which provide very important and necessary foundations on which applications ride.

The chapter then presented application-specific performance requirements or objectives compiled by MEF from variety of sources in public domain. These APOs are then mapped to MEF-defined standard CPOs and PTs. This mapping is crucial to standardizing CE services particularly in peering CENs. The mapping was also described in this chapter.

The chapter then covered the applicable CE services including E-Access service for peering CENs to meet the network functionality needed by customer applications. Examples of network functionality include IP backhaul, mobile backhaul, streaming and switched video transport, site-to-site connectivity, connection for cloud computing services, and network connectivity for emerging applications such as IoT, CPS, and VR.

The chapter next briefly described a process to convert information about customer applications and topology into a design for a CE service based on CENs and peering CENs. Finally, the chapter provided transitions to next steps needed in further enhancing peering of CENs to accommodate emerging trends which is covered in some detail in the final chapter of this book.

9

NEXT STEPS IN PEERING CARRIER ETHERNET NETWORKS

Now this is not the end. It is not even the beginning of the end. But it is, perhaps, the end of the beginning.

Remarks by Winston Churchill

Changes on the horizon in information and communication technology are sure to cause, in next few years, profound impact on current state. We learned from all previous chapters in this book that change is one thing that is constant in this field. Starting with the invention of the telephone, these changes have come at regular intervals. These included inventions of switching and multiplexing, development of Nyquist theorem, and Shannon's information theory, C/C++, and UNIX, inventions of transistors, microprocessors, lasers, and fiber optics, development of TCP/IP, Ethernet, OSI-7 layer model, concept of open source systems, Internet, geography-centric blueprint of Internet architecture, xDSL, DOCSIS specification, move from circuit switching to packet switching, and so forth. Of course, as described earlier in Chapter 1, these changes were accompanied with changes in regulatory and business landscapes, leading to a dynamic situation. These changes, though seemingly chaotic, have propelled the technology toward digitization of all services and making them available globally, hardly imaginable just a few decades ago.

Years starting from early 1990s have been particularly tumultuous because of the fast clip at which changes arrived due to Web browsers, World Wide Web, wireless communication, smartphones, VoIP, high-definition TVs, and smart TVs. These changes led to the emergence of data network as the most dominant network and the adoption of Ethernet as the most popular protocol at layer 2, with the result that over 90% of LAN traffic around the globe is on Ethernet today. Subsequent enhancements in Ethernet led to Ethernet over DWDM at layer 1 which extended the reach of Ethernet, covering large distances beyond LAN as well.

Peering Carrier Ethernet Networks. http://dx.doi.org/10.1016/B978-0-12-805319-5.00009-5

In Chapter 4, we covered how the communication industries came together to form MEF to leverage benefits of Ethernet technology by defining and standardizing carrier Ethernet (CE) services, making them reliable and scalable and made them carrier grade by specifying QoS and service management. Carrier Ethernet is a marketing term for these carrier-grade Ethernet services and associated networks on which these services provided are known as carrier Ethernet networks (CENs).

In Chapter 5, we covered that the growth in CE services is leading to the expansion of these services beyond one operator's CEN. This expansion is also coupled with the fact that today subscribers, particularly business subscribers, have many locations that are not all in one operator's footprint. These reasons are making peering of CENs necessary. To address this necessity, MEF defined an E-access service type.

Now, as we discussed in the last chapter, there are trends that are emerging, which are bound to cause profound impacts on the current state of information and communication technology landscape. These emerging trends are summarized in the next section.

9.1 Emerging Trends

In the last chapter, we covered that many new subscriber applications are emerging including cyber-physical systems (CPS), Internet of things (IoT), robotics, virtual reality, TVs with high definition, ultra-high definition, and 3D video capabilities. These applications will lead to more immersive, pervasive, and responsive customer experience and to a smart connected planet. They also will be requiring higher bandwidth, low frame delay, low frame delay variation, and low frame loss probability. Also, the introduction of 5G wireless communications in the coming years will require higher bandwidth in mobile backhaul. More widespread deployment of VoIP and phasing out of legacy voice will also require better performance (lower frame delay, frame delay variation, and frame loss probability). Phasing out of TDM networks will funnel additional traffic to data networks. In addition to these new applications, there is a push by operators to introduce network function virtualization (NFV) and software-defined networks (SDN) to reduce Capex and Opex as well as to encourage customer driven on-demand processes for ordering and configuring services. The new subscriber applications and network virtualization are going to make clouds grow rapidly, further making demands on higher bandwidth, low frame delay, low frame delay variation, and low frame loss probability.

The Ethernet technology is uniquely suited to meet these requirements. Additionally, because today over 90% of LAN traffic around the globe is over Ethernet, extending this technology over MAN, RAN, and WAN will eliminate multiple protocol translations leading to reduction in Capex and Opex. These factors will place CENs and peering CENs right in the front and center. In the next section, we will examine some of the next steps needed in the Ethernet technology and peering CENs to meet the growing demands.

9.2 Next Steps in Ethernet Technology and Peering Carrier Ethernet Networks

Implementation of Ethernet technology–based CENs and peering CENs is essential to meeting the demands of higher bandwidth and performance in IP backhaul, mobile backhaul, streaming and switched video transport, site-to-site connectivity, and network connectivity for cloud computing services for emerging applications. As this implementation grows, it will require further increase in bandwidth especially in the aggregation networks and so will need accelerated widespread deployment of 100 Gbps Ethernet technology. As deployments of 100 Gbps Ethernet technology along with the deployments of emerging applications grow, these in turn will create demand for even higher bandwidths and performance. That will accelerate implementation and deployment of 400 Gbps Ethernet technology. That will, in turn, push for even higher bandwidths beyond 400 Gbps, for example, for developing Ethernet technology for 1 Tbps and beyond.

A comparison[37,38] of PBB-TE and MPLS-TP was presented in Section 5.1 of Chapter 5. It is important to recall that the PBB-TE was standardized by the IEEE as a connection-oriented transport enhancement that comes from the Ethernet world of IEEE 802.1Q, and the MPLS-TP was standardized jointly by the ITU-T and IETF having its origins in the MPLS (layer 2.5) world always requiring a foreign layer 2, which is increasingly Ethernet, to transport it. Both provide the same transport functionality for CE services but are very different. A study[45] presented at the MPLS and Ethernet World Congress, 2012, compared the IEEE 802.1Q-based Ethernet protocol with the MPLS-TP protocol. The comparison was based on 10 functionalities including fault management, performance management, automatic protection switching (APS) mechanism, QoS management, handling diverse traffic types, high-accuracy timing distribution, integration with networks, Capex, Opex, and security. Ethernet protocol got 89 points compared to 52 for MPLS-TP protocol. Despite this, the industry acceptance of PBB-TE has so far turned out to be low, although several vendors still support PBB-TE technology in their hardware. The acceptance of MPLS-TP is also not growing. In the meanwhile, many off-the-shelf switch chip vendors have implemented PBB-TE, PBB, IP, MPLS, and MPLS-TP protocols, and as we discussed in Section 3.3 of Chapter 3, these chips are capable of processing all these protocols at line rate. Although the debates about PBB-TE and MPLS-TP still continue, focus has moved back toward mainline MPLS because it already has a large installed base. However, this area is still evolving and will be focus of attention in the future due to increasing implementation of "switch many, route once" methodology in data networks and the growing adoption of CENs and peering CENs and also due to the emerging new applications requiring higher bandwidth, performance, and scalability, and better separation between user and network operators.

There is also a need for closing gaps between MEF-defined attributes/parameters and ATIS-defined ASR fields/values as discussed in Chapter 6. For example, ASR still does not have a field to specify performance tier which is part of the normative performance service attribute per MEF 26.1 and MEF 26.2 specifications. Sometimes these gaps lead to delays in implementation of peering CENs.

Another important area that needs examination is about alternate models of ENNI, for example, use of shared ENNI instead of each operator ordering their own ENNI. This is because as the bandwidth increases, shared ENNI will utilize the capacity of the ENNI link between operators more efficiently compared to each operator having their own ENNIs.

With the growth of peering CENs, there is a strong need for implementation and governance guidelines for ENNIs. TDM-based meet points have been around for a long time, and therefore, there are standards for coordination and governance of these TDM-based meet points. Similar structure and organization are needed for ENNIs for peering CENs. MEF has initiated efforts in this direction with the Ethernet interconnect points project[46]. In addition to this, some MEF specifications are getting revised, and new ones will be coming out in the future in support of CE services in CENs and peering CENs.

Current implementation of CENs and peering CENs are based on physical network function. As operators move toward SDN and NFV technologies, CENs and peering CENs will need to embrace virtual network function too. To address this MEF has initiated, in 2016, a Third Network vision and a Lifecycle Service Orchestration (LSO) reference architecture[36].

Current SOAM for peering CENs is based on limited performance management and fault management between operators. Enhancing this to include additional performance management,

SLA management, and trouble ticket management areas as well between operators is worth looking into to further improve SOAM implementation for peering CENs.

As discussed in Chapter 8, presently, CoS performance objectives (CPOs) are for point-to-point and multipoint-to-multipoint cases. They have to be developed for rooted multipoint cases as well. Additionally, some CPOs are missing for many applications including emerging applications. This gap needs to be filled.

Changes in regulatory environment are needed too. Current regulatory regime does not have level playing field for telecommunications companies on one hand and MSOs on the other hand. Also, these current regulations need to be revisited in view of rapid conversion of voice into another data application i.e., VoIP. This is due to the fact that traditional voice is regulated but data is unregulated. Another regulatory area to watch is about net neutrality. The net neutrality rules were created by the Federal Communications Commission in early 2015. A three-judge panel at the US Court of Appeals for the District of Columbia Circuit, on June 14, 2016, upheld, by a two-to-one decision, the Federal Communications Commission net neutrality rules prohibiting broadband companies from blocking or slowing the delivery of internet content to consumers. Large content providers support net neutrality rules, whereas large MSOs and telecommunications operators oppose it and have promised to take the case to the US Supreme Court.

9.3 Next Steps in Operations and Business Support Systems

As we know that in addition to networks, the operations and business support systems (OSS/BSS) provide an important foundation for subscriber applications and are critical to the operations of service providing and access providing operators. Current OSS/BSS systems are based on legacy systems with many silos. These large and complex IT systems have to be enhanced and modified to work with newer applications and newer technologies including CENs, peering CENs, SDN, NFV, and cloud. Going forward, the OSS/BSS systems will be required to provision applications based on new technologies and handle real-time billing. This transition will create some interesting challenges because these legacy systems are well entrenched and many legacy OSS/BSS systems, for example, use billing account numbers as primary key for provisioning therefore changing billing systems will impact provisioning systems as well. In addition, in the transition period, OSS/BSS systems will have to deal with both physical network function–based and virtual network function–based CENs and peering CENs. All these will make the situation even more complicated before it gets better. In closing, although we have come to the end of this book, the journey is just beginning!

9.4 Chapter Summary

We now have come to the end of this book, journeying through the changing landscape beginning with the invention of telephone in 1876 to the present state of information and communication technology in 2016 where peering CENs are poised to play an important role in IP backhaul, mobile backhaul, streaming, and switched video transport, site-to-site connectivity, and connections for cloud computing services to facilitate network connectivity for existing as well as emerging applications. In this journey, we saw how Ethernet evolved as the most adopted protocol in layer 2 resulting in over 90% of LAN traffic around the globe on Ethernet today. Further enhancements in Ethernet led to Ethernet over DWDM at layer 1 which extended its reach beyond LAN to MAN, RAN, and WAN.

In this journey, we also saw that to leverage benefits of Ethernet technology including higher bandwidth, low frame delay, low frame delay variation, and low frame loss probability as well as elimination of multiple protocol translations, MEF-defined and standardized CE services, making them reliable, scalable, and carrier grade resulting in CENs. The expansion of CENs has now necessitated peering of CENs. Demands of higher bandwidth and performance by emerging applications on data networks and the push by operators to lower Capex and Opex are propelling CENs and peering CENs right in the front and center.

In this chapter, we examined some of the next steps needed in the Ethernet technology and peering CENs and also in OSS/BSS to meet the growing demands because these are critical to subscriber applications and to the operations of service providing and access providing operators.

REFERENCES

1. Alexander Graham Bell, granted US patent number 174,465 for "improvements in telegraphy", March 7, 1876.
2. Almon Strowger granted US patent number 0,447,918 for "automatic telephone exchange", March 10, 1891.
3. Squirer MGO. *The invention of "Wired Wireless" telephony and telegraphy, extract from the laboratory note book.* Signal Corps Laboratory; September 30, 1910.
4. Major George O. Squirer, granted US patent number 980,356;980,357;980,358 and 980,359 for "multiplex telephony and telegraphy", January 3, 1911.
5. Nyquist H. Certain topics in Telegraph Transmission Theory. *Physical Reviews* 1928;**32**:110.
6. Shannon EC. A Mathematical Theory of Communication. *The Bell System Technical Journal* July–October, 1948;**27**:379–423.
7. Kleinrock L. *Information flow in large communication nets, RLE quarterly progress report.* Massachusetts Institute of Technology; April 1962.
8. Baran P. *On distributed communications*, vol. I–XI. August 1964. RAND report RM 3420.
9. Davies DW, Bartlett KA, Scantlebury RA, Wilkinson PT. A digital communications network for computers giving rapid response at remote terminals. *ACM Symposium on Operating Systems Problems* October 1967.
10. Cerf V, Kahn R. A protocol for packet network intercommunication. *IEEE Transactions on Communications* May 1974;**22**(5).
11. Metcalfe R, Boggs D. Ethernet: distributed packet-switching for local computer networks. *Communications of the ACM* July 1976;**19**(Issue 7):395–405.
12. Mockapetris P. Domain names – concepts and facilities. *RFC#* November 1983;**882**.
13. Tim-Berners L. *WWW – software in public domain.* CERN; April 30, 1993.
14. Kangovi S, granted US patent numbers 7,912,195, March 22, 2011; 8,594,289, November 26, 2013 and 9,070,154, June 30, 2015, for "method for provisioning subscribers, products, and services in a broadband network". There is also an International patent application (PCT/US2007/070591) published by World Intellectual Property Organization (WIPO) under WO/2007/143712A3.
15. Maxwell JC. *A Dynamical Theory of the Electromagnetic Field, Presented at the Royal Society of London.* December 8, 1864.
16. Hertz HR. Ueber die Ausbreitungsgeschwindigkeit der electrodynamischen Wirkungen. *Annalen der Physik* May 1888;**270**(7):551–69.
17. Marconi G, granted British patent number 12,039 for "Improvements in transmitting electrical impulses and signals and in apparatus there-for", June 2, 1896. Complete specification filed on March 2, 1897 and accepted, July 2, 1897.
18. Pouzin L. Presentation and major design aspects of the CYCLADES network. In: *Data communications symposium, Tampa, Florida.* November 1973. p. 80–5.
19. Despres LG. A packet network with graceful saturated operation. In: *Proceedings of ICCC, Washington, DC.* October 1972. p. 345–51.
20. *TELENET application to "Institute and operate a public packet switched data network in the United States.* April 1974. Filed with FCC in October 1973 and approved by FCC by Order 46 FCC2d 680, File P-C-8750.
21. Belton RC, Thomas JR. *The UKPO packet switching experiment.* ISS Munich; 1974.
22. Danet A, Depres R, LesRest A, Pichon G, Ritzenthaler S. The French public packet switching service: the TRANSPAC network. In: *Proceedings of ICCC, Toronto.* August 1976. p. 251–60.
23. Clipsham WW, Glave FE, Narraway ML. DATAPAC network overview. In: *Proceedings of ICCC, Toronto.* August 1976. p. 131–6.
24. Nakamura R, Ishino F, Sasakoka M, Nakamura M. Some design aspects of a Public Switched Network. In: *Proceedings of ICCC, Toronto.* August 1976. p. 317–22.
25. http://www.ieee802.org/3/.
26. http://www.itu.int/rec/T-REC-X.200.
27. Management of NSFNET. Hearing before the subcommittee on science of the committee on science, space, and technology, US House of representatives. In: *102nd Congress, Second session, Transcript of March 12, 1992.*
28. Cioffi J. *The need for speed.* The Great Minds, Great Ideas Project, EE Times 1979www.eetimes.com/disruption/essays/cioffi.jhtml.
29. ANSI standard T1.413 for Discrete Multi-tone (DMT) line-coding scheme for ADSL; 1998.

30. Kao C, Hockham G. Circuit waveguides for optical frequencies. *Proceedings of IEEE* 1966;**113**: 1151–8.

31. Maurer RD, Schultz PC, granted US patent number 3,659,915 for "fused silica optical waveguide", May 11, 1970.

32. Digital, Intel and Xerox Specification. The ethernet – a local area network data link layer and physical layer specifications. *DIX Specif AA-K759B-TK* November 1982;**2**. 0.

33. Boggs DR, Mogul JC, Kent CA. *Measured capacity of an ethernet: myths and reality.* Western Research Laboratory Research Report 88/4. Digital Equipment Corporation; September 1988.

34. http://www.ieee802.org/.

35. https://www.itu.int/rec/T-REC-Y.1731/en.

36. http://www.mef.net.

37. https://www.cse.iitb.ac.in/internal/techreports/reports/TR-CSE-2008-11.pdf; 2008.

38. https://en.wikipedia.org/wiki/Provider_Backbone_Bridge_Traffic_Engineering#cite_note-4; 2016.

39. www.atis.org/01_committ_forums/OBF/documents.asp.

40. http://www.iconnective.com.

41. http://www.itu.int/rec/T-REC-M/e.

42. http://www.tmforum.org.

43. http://www.etsi.org/technologies-clusters/technologies/nfv.

44. National Science Foundation Cyber-Physical Systems (CPS) – NSF; 2016.

45. Stein JY. Comparing access packet-based technologies. In: *Presented at the MPLS & ethernet congress – 2012, Paris.* February 9, 2012.

46. www.mef.net/eipproject; 2016.

INDEX

Note: 'Page numbers followed by "f" indicate figures and "t" indicate tables.'

Printed in the United States
By Bookmasters